CAMBRIDGE LIBRARY COLLECTION

Books of enduring scholarly value

Physical Sciences

From ancient times, humans have tried to understand the workings of the world around them. The roots of modern physical science go back to the very earliest mechanical devices such as levers and rollers, the mixing of paints and dyes, and the importance of the heavenly bodies in early religious observance and navigation. The physical sciences as we know them today began to emerge as independent academic subjects during the early modern period, in the work of Newton and other 'natural philosophers', and numerous sub-disciplines developed during the centuries that followed. This part of the Cambridge Library Collection is devoted to landmark publications in this area which will be of interest to historians of science concerned with individual scientists, particular discoveries, and advances in scientific method, or with the establishment and development of scientific institutions around the world.

Michael Faraday

Encouraged to share his memories of Michael Faraday (1791–1867), John Hall Gladstone (1827–1902) published in 1872 this short work about his late friend's life and career. Faraday's successor as Fullerian Professor of Chemistry at the Royal Institution, Gladstone discusses how Faraday approached science, and the value of his discoveries. Offering informed insights into Faraday's character, Gladstone includes a number of extracts from personal letters. The work also includes a translation of part of the eulogy given by Jean-Baptiste Dumas at the Académie des Sciences, as well as an anonymous poem honouring Faraday and published in *Punch* shortly after his death. An appendix lists the numerous learned societies to which Faraday belonged. Also reissued in this series are *The Life and Letters of Faraday* (1870), compiled by Henry Bence Jones, and John Tyndall's *Faraday as a Discoverer* (1868).

Michael Faraday

JOHN HALL GLADSTONE

CAMBRIDGE
UNIVERSITY PRESS

University Printing House, Cambridge, CB2 8BS, United Kingdom

Cambridge University Press is part of the University of Cambridge.
It furthers the University's mission by disseminating knowledge in the pursuit of
education, learning and research at the highest international levels of excellence.

www.cambridge.org
Information on this title: www.cambridge.org/9781108070096

© in this compilation Cambridge University Press 2014

This edition first published 1872
This digitally printed version 2014

ISBN 978-1-108-07009-6 Paperback

MICHAEL FARADAY.

MICHAEL FARADAY.

BY

J. H. GLADSTONE, Ph.D., F.R.S.

London:
MACMILLAN AND CO.
1872.

LONDON :
R. CLAY, SONS, AND TAYLOR, PRINTERS,
BREAD STREET HILL.

PREFACE.

SHORTLY after the death of Michael Faraday, Professor
Auguste de la Rive, and others of his friends, gave to the
world their impressions of his life, his character, and his
work ; Professor Tyndall drew his portrait as a man of
science ; and after a while Dr. Bence Jones published his
biography in two octavo volumes, with copious extracts
from his journals and correspondence.   In a review of
this "Life and Letters" I happened to mention my thought
of giving to the public some day my own reminiscences
of the great philosopher; several friends urged me to do
so, not in the pages of a magazine, but in the form of a
little book designed for those of his fellow-countrymen
who venerate his noble character without being able to
follow his scientific researches.   I accepted the task.
Professor Tyndall and Dr. Bence Jones, with Messrs.
Longman, the publishers, kindly permitted me to make
free use of their materials ; but I am indebted to the

Corporation of the Trinity House, and to many friends, for a good deal of additional information ; and in compiling my book I have preferred, where practicable, to illustrate the character of Faraday by documents or incidents hitherto unpublished, or contained in those sketches of the philosopher which are less generally known.

It is due to myself to say that I had pretty well sketched out the second part of this book before I read M. Dumas' "Eloge Historique." The close similarity of my analysis of Professor Faraday's character with that of the illustrious French chemist may perhaps be accepted as an additional warrant for the correctness of our independent estimates.

CONTENTS.

MICHAEL FARADAY.

SECTION I.

THE STORY OF HIS LIFE.

AT the beginning of this century, in the neighbourhood
of Manchester Square, London, there was an inquisitive
boy running about, playing at marbles, and minding his
baby-sister. He lived in Jacob's Well Mews, close by,
and was learning the three R's at a common day-school.
Few passers-by would have noticed him, and none certainly
would have imagined that this boy, as he grew up, was to
achieve the truest success in life, and to die honoured by
the great, the wise, and the good. Yet so it was ; and to
tell the story of his life, to trace the sources of this success,
and to depict some of the noble results of his work, are the
objects of this biographical sketch.

It was not at Jacob's Well Mews, but in Newington Butts,
that the boy had been born, on September 22, 1791, and his
parents, James and Margaret Faraday, had given this, their
third child, the unusual name of Michael. The father was
a journeyman blacksmith, and, in spite of poverty and feeble

B

health, he strove to bring up his children in habits of industry and the love of God.

Of course young Michael must soon do something for his living. There happened to be a bookseller's shop in Bland-ford Street, a few doors from the entrance to the Mews, kept by a Mr. Riebau, an intelligent man, who is said to have had a leaning to astrology; and there he went as errand boy when thirteen years old. Many a weary walk he had, carrying round newspapers to his master's customers; but he did his work faithfully; and so, after a twelvemonth, the bookseller was willing to take him as an apprentice, and that without a premium.

Now, a boy in a bookseller's shop can look at the inside as well as the outside of the books he handles, and young Faraday took advantage of his position, and fed on such intellectual food as Watts's " Improvement of the Mind," Mrs. Marcet's " Conversations on Chemistry," and the article on "Electricity" in the *Encyclopædia Britannica*, besides such lighter dishes as Miss Burney's "Evelina;" nor can we doubt that when he was binding Lyons' " Experiments on Elec-tricity," and Boyle's "Notes about the Producibleness of Chymicall Principles," he looked beyond the covers.[1]   And

---

[1] These books, with others bound by Faraday, are preserved in a special cabinet at the Royal Institution, together with more valuable documents,—the laboratory notes of Davy and those of Faraday, his notes of Tatum's and Davy's lectures, copies of his published papers with annotations and indices, notes for lectures and Friday evening dis-courses, account books and various memoranda, together with letters from Wollaston, Young, Herschel, Whewell, Mitscherlich, and many others of his fellow-workers in science. These were the gift of his widow, in accordance with his own desire.

his thirst for knowledge did not stop with reading : he must see whether Mrs. Marcet's statements were correct, and so, to quote his own words, " I made such simple experiments in chemistry as could be defrayed in their expense by a few pence per week, and also constructed an electrical machine, first with a glass phial, and afterwards with a real cylinder, as well as other electrical apparatus of a corresponding kind."

One day, walking somewhere in the neighbourhood of Fleet Street, he saw in a shop-window a bill announcing that lectures on natural philosophy were delivered by Mr. Tatum, at 53, Dorset Street, at eight in the evening, price of admission one shilling. He wanted to hear these lectures. His master's permission was obtained, but where was the money to come from? The needful shillings were given him by his elder brother, Robert, who earned them as a blacksmith; and so Michael Faraday made his first acquaintance with scientific lectures. And not with lectures only, for Tatum's house was frequented by other earnest students, and lifelong friendships were formed. Among these students was Benjamin Abbott, a young Quaker, who had received a good education, and had then a situation in a City house as confidential clerk. With him Faraday chatted on philosophy or anything else, and happily for us he chatted on paper, in letters of that fulness and length which the penny post and the telegraph have well-nigh driven out of existence ; and happily for us, too, Abbott kept those letters, and Dr. Bence Jones has published them. They are wonderful letters for a poor bookseller's apprentice ; they bear the stamp of an innate gentleman and philosopher.

Long afterwards, when Benjamin Abbott was an old man,

he used to tell how Faraday made his first experiments in
the kitchen of his house, and delivered his first lecture from
the end of that kitchen table.   The electrical machine
made by him in those early days came into the possession
of Sir James South, and now forms one of the treasures of
the Royal Institution.

As the eager student drank in the lectures of Tatum, he
took notes, and he afterwards wrote them out carefully in a
clear hand, numbering and describing the different experiments that he saw performed, and making wonderfully neat
drawings of the apparatus, in good perspective.   These
notes he bound in four volumes, adding to each a copious
index, and prefixing to the first this dedication to his
master :—

" To Mr. G. Riebau.

"Sir,

        "When first I evinced a predilection for the sciences,
but more particularly for that one denominated electricity,
you kindly interested yourself in the progress I made in the
knowledge of facts relating to the different theories in existence, readily permitting me to examine those books in your
possession that were in any way related to the subjects then
occupying my attention.   To you, therefore, is to be attributed the rise and existence of that small portion of knowledge relating to the sciences which I possess, and accordingly to you are due my acknowledgments.

        " Unused to the arts of flattery, I can only express my
obligations in a plain but sincere way.   Permit me, therefore, Sir, to return thanks in this manner for the many

favours I have received at your hands and by your means,
and believe me,

"Your grateful and obedient Servant,

"M. FARADAY."

Now there happened to be lodging at Mr. Riebau's a
notable foreigner of the name of Masquerier. He was a
distinguished artist, who had painted Napoleon's portrait,
and had passed through the stirring events of the first
French Revolution, not without serious personal danger,
and was now finding a refuge and a home in London. He
was struck with the intelligence of the apprentice, whose
duty it was to do various offices for him; and he lent the
young man his books, and taught him how to make the
drawings in perspective which have already been alluded to.

But the lectures in Dorset Street were not the only ones
that Michael Faraday attended; and as the Royal Institution is the central scene of all his subsequent history, we
must pay a mental visit to that building. Turning from the
busy stream of Piccadilly into the quiet of Albemarle
Street, we see, in a line with the other houses, a large
Grecian façade with fourteen lofty pilasters. Between these
are folding doors, which are pushed open from time to time
by grave-looking gentlemen, many of them white-headed;
but often of an afternoon, and always on Friday evening
during the season, the quiet street is thronged with carriages
and pedestrians, ladies and gentlemen, who flock through
these folding doors. Entering with them, we find ourselves
in a vestibule, with a large stone staircase in front, and rooms
opening on the right and left. The walls of these rooms

are lined with myriads of books, and the tables are covered
with scientific and other periodicals of the day, and there
are cabinets of philosophical apparatus and a small museum.
Going up the broad staircase and turning to the right, we
pass through an ante-room to the lecture theatre. There
stands the large table, horseshoe-shaped, with the necessary
appliances for experiments, and behind it a furnace and
arrangements for black-board and diagrams ; while round
the table as a centre range semicircular seats, rising tier
above tier, and surmounted by a semicircular gallery, the
whole capable of seating 700 persons. On the basement
is a new chemical laboratory, fitted up with modern appliances, and beyond it the old laboratory, with its furnaces
and sand-bath, its working tables and well-stored shelves,
flanked by cellars that look like dark lumber-rooms.   A
narrow private staircase leads up to the suite of apartments
in which resides the Director of the house. Such is the
Royal Institution of Great Britain, incorporated by Royal
Charter in the year 1800, " for the diffusing knowledge and
facilitating the general introduction of useful mechanical inventions and improvements, and for teaching, by courses of
philosophical lectures and experiments, the application of
science to the common purposes of life ;"—with the motto,
" Illustrans commoda vitæ."   Fifty or sixty years ago the
building was essentially what it is now, except the façade
and entrance, and that the laboratory, which was considered
a model of perfection, was even darker than at present, and
in the place of the modern chemical room there was a
small theatre. The side room, too, was fitted up for actual
work, though even at mid-day it had to be artificially

lighted; and beyond this there was, and still is, a place
called the Froggery, from a certain old tradition of frogs
having been kept there. The first intention of the founders
to exhibit useful inventions had not been found very practicable, but the place was already famous with the memories
of Rumford and Young; and at that time the genius of Sir
Humphry Davy was entrancing the intellectual world with
brilliant discoveries, and drawing fashionable audiences to
Albemarle Street to listen to his eloquent expositions.

Among the customers of the bookseller in Blandford
Street was a Mr. Dance, who, being a member of the Royal
Institution, took young Faraday to hear the last four public
lectures of Davy. The eager student sat in the gallery, just
over the clock, and took copious notes of the Professor's
explanations of radiant matter, chlorine, simple inflam-
mables, and metals, while he watched the experiments that
were performed. Afterwards he wrote the lectures fairly out
in a quarto volume, that is still preserved—first the theoretical portions, then the experiments with drawings, and
finally an index. "The desire to be engaged in scientific
occupation, even though of the lowest kind, induced me,"
he says, "whilst an apprentice, to write, in my ignorance of
the world and simplicity of my mind, to Sir Joseph Banks,
then President of the Royal Society.   Naturally enough,
'No answer' was the reply left with the porter."

On the 7th of October his apprenticeship expired, and on
the next day he became a journeyman bookbinder under a
disagreeable master—who, like his friend the artist, was a
French *émigré*. No wonder he sighed still more for congenial occupation.

Towards the end of that same October Sir Humphry Davy was working on a new liquid which was violently explosive, now known as chloride of nitrogen,—and he met with an accident that seriously injured his eye, and produced an attack of inflammation. Of course, for a while he could not write, and, probably through the introduction of M. Masquerier,[1] the young bookseller was employed as his amanuensis. This, however, Faraday himself tells us lasted only "some days;" and in writing years afterwards to Dr. Paris, he says, "My desire to escape from trade, which I thought vicious and selfish, and to enter into the service of Science, which I imagined made its pursuers amiable and liberal, induced me at last to take the bold and simple step of writing to Sir H. Davy, expressing my wishes, and a hope that, if an opportunity came in his way, he would favour my views; at the same time I sent the notes I had taken of his lectures." Davy, it seems, called with the letter on one of his friends—at that time honorary inspector of the models and apparatus—and said, "Pepys, what am I to do? Here is a letter from a young man named Faraday; he has been attending my lectures, and wants me to give him employment at the Royal Institution—*what can I do?*" "Do?" replied Pepys; "put him to wash bottles: if he is good for anything, he will do it directly; if he refuses, he is good for nothing." "No, no," replied Davy, "we must try him with something better than that."

So Davy wrote a kind reply, and had an interview with the young man upon the subject; in which, however, he advised

[1] This seems probable from some remarks of Faraday to Lady Burdett Coutts.

him to stick to his business, telling him that "Science was a harsh mistress, and, in a pecuniary point of view, but poorly rewarding those who devoted themselves to her service." He promised him the work of the Institution, and his own besides.

But shortly afterwards the laboratory assistant was discharged for misconduct, and so it happened that one night the inhabitants of quiet Weymouth Street were startled by the unusual apparition of a grand carriage with a footman, which drew up before the house where Faraday lived, when the servant left a note from Sir Humphry Davy. The next morning there was an interview, which resulted in the young aspirant for scientific work being engaged to help the famous philosopher. His engagement dates from March 1, 1813, and he was to get 25*s.* per week, and a room in the house. The duties had been previously laid down by the managers :—
"To attend and assist the lecturers and professors in preparing for, and during lectures. Where any instruments or apparatus may be required, to attend to their careful removal from the model room and laboratory to the lecture-room, and to clean and replace them after being used, reporting to the managers such accidents as shall require repair, a constant diary being kept by him for that purpose. That in one day in each week he be employed in keeping clean the models in the repository, and that all the instruments in the glass cases be cleaned and dusted at least once within a month."

The young assistant did not confine himself to the mere discharge of these somewhat menial duties. He put in order the mineralogical collection; and from the first we find him occupying a higher position than the minute quoted above would indicate.

In the course of a few days he was extracting sugar from beet-root; but all his laboratory proceedings were not so pleasant or so innocent as that, for he had to make one of the worst smelling of all chemical compounds, bisulphide of carbon; and as Davy continued to work on the explosive chloride of nitrogen, his assistant's career stood some chance of being suddenly cut short at its commencement. Indeed, it seems that before the middle of April he had run the gauntlet of four separate explosions. Knowing that the liquid would go off on the slightest provocation, the experimenters wore masks of glass, but this did not save them from injury. In one case Faraday was holding a small tube containing a few grains of it between his finger and thumb, and brought a piece of warm cement near it, when he was suddenly stunned, and on returning to consciousness found himself standing with his hand in the same position, but torn by the shattered tube, and the glass of his mask even cut by the projected fragments. Nor was it easy to say when the compound could be relied on, for it seemed very capricious; for instance, one day it rose quietly in vapour in a tube exhausted by the air-pump, but the next day, when subjected to the same treatment, it exploded with a fearful noise, and Sir Humphry was cut about the chin, and was struck with violence on the forehead. This seems to have put an end to the experiments.

Nevertheless, in spite of disagreeables and dangers, the embryo philosopher worked on with a joyful heart, beguiling himself occasionally with a song, and in the evening playing tunes on his flute.

The change in Michael Faraday's employment naturally

made him more earnest still in the pursuit of knowledge.
He was admitted as a member of the " City Philosophica
Society," a fraternity of thirty or forty men in the middle
or lower ranks of life, who met every Wednesday evening
for mutual instruction; and here is a contemporary picture
of him at one of its debates :—

> " But hark !   A voice arises near the chair !
> Its liquid sounds glide smoothly through the air ;
> The listening muse with rapture bends to view
> The place of speaking, and the speaker too.
> Neat was the youth in dress, in person plain ;
> His eye read thus, *Philosopher in grain ;*
> Of understanding clear, reflection deep ;
> Expert to apprehend, and strong to keep.
> His watchful mind no subject can elude,
> Nor specious arts of sophists e'er delude ;
> His powers, unshackled, range from pole to pole :
> His mind from error free, from guilt his soul.
> Warmth in his heart, good humour in his face,
> A friend to mirth, but foe to vile grimace ;
> A temper candid, manners unassuming,
> Always correct, yet always unpresuming.
> Such was the youth, the chief of all the band ;
> His name well known, Sir Humphry's right hand.
> With manly ease towards the chair he bends,
> With Watts's Logic at his finger-ends."

Another way in which he strove to educate himself is
thus described in his own words :—" During this spring
Magrath and I established the mutual improvement plan,
and met at my rooms up in the attics of the Royal Institution, or at Wood Street at his warehouse.   It consisted,
perhaps, of half-a-dozen persons, chiefly from the City
Philosophical Society, who met of an evening to read

together, and to criticise, correct, and improve each other's pronunciation and construction of language. The discipline was very sturdy, the remarks very plain and open, and the results most valuable. This continued for several years."

Seven months after his appointment there began a new passage in Faraday's life, which gave a fresh impulse to his mental activity, and largely extended his knowledge of men and things. Sir Humphry Davy, wishing to travel on the Continent, and having received a special pass from the Emperor Napoleon, offered to take him as his amanuensis : he accepted the proposal, and for a year and a half they wandered about France, Italy, and Switzerland, and then they returned rapidly by the Tyrol, Germany, and Holland.

From letters written when abroad we can catch some of the impressions made on his mind by these novel scenes. " I have not forgot," he writes to Abbott, " and never shall forget, the ideas that were forced on my mind in the first days. To me, who had lived all my days of remembrance in London, a city surrounded by a flat green country, a hill was a mountain, and a stone a rock; for though I had abstract ideas of the things, and could say rock and mountain, and would talk of them, yet I had no perfect ideas. Conceive then the astonishment, the pleasure, and the information which entered my mind in the varied county of Devonshire, where the foundations of the earth were first exposed to my view, and where I first saw granite, limestone, &c., in those places and in those forms where the ever-working and all-wonderful hand of nature had placed them. Mr. Ben., it is impossible you can conceive my feelings, and it is as impossible for me to describe them.   The sea then presented a new source of information and interest; and on approaching the shore of France, with what eagerness, and how often, were my eyes directed to the South!   When arrived there, I thought myself in an uncivilized country; for never before nor since have I seen such wretched beings as at Morlaix."   His impression of the people was not improved by the fact of their having arrested the travellers on landing, and having detained them for five days until they had sent to Paris for verification of their papers.

Again, to her towards whom his heart was wont to turn from distant lands with no small longing: "I have said nothing as yet to you, dear mother, about our past journey, which has been as pleasant and agreeable (a few things excepted, in reality nothing) as it was possible to be.   Sir H. Davy's high name at Paris gave us free admission into all parts of the French dominions, and our passports were granted with the utmost readiness.   We first went to Paris, and stopped there two months; afterwards we passed, in a southerly direction, through France to Montpellier, on the borders of the Mediterranean.   From thence we went to Nice, stopping a day or two at Aix on our way; and from Nice we crossed the Alps to Turin, in Piedmont.   From Turin we proceeded to Genoa, which place we left afterwards in an open boat, and proceeded by sea towards Lerici.   This place we reached after a very disagreeable passage, and not without apprehensions of being overset by the way.   As there was nothing there very enticing, we continued our route to Florence; and, after a stay of three weeks or a month, left that fine city, and in four days arrived here at

Rome. Being now in the midst of things curious and interesting, something arises every day which calls for attention
and observations. The relics of ancient Roman magni-
ficence, the grandeur of the churches, and their richness
also—the difference of habits and customs, each in turn
engages the mind, and keeps it continually employed.
Florence, too, was not destitute of its attractions for me,
and in the Academy del Cimento and the museum attached
to it is contained an inexhaustible fund of entertainment
and improvement; indeed, during the whole journey, new
and instructive things have been continually presented to
me.   Tell B. I have crossed the Alps and the Apennines;
I have been at the Jardin des Plantes; at the museum
arranged by Buffon ; at the Louvre, among the *chefs d'œuvre*
of sculpture and the masterpieces of painting; at the
Luxembourg Palace, amongst Rubens' works ; that I have
seen a GLOWWORM ! ! ! waterspouts, torpedo, the museum at
the Academy del Cimento, as well as St. Peter's, and some
of the antiquities here, and a vast variety of things far too
numerous to enumerate."

But he kept a lengthy journal, and as we turn over the
pages—for the best part of it is printed by Bence Jones—we
meet vivid sketches of the provokingly slow custom-house
officers, the postilion in jack-boots, and the thin pigs of
Morlaix—pictures of Paris, too, when every Frenchman was
to him an unintelligible enemy; when the Apollo Belvidere,
the Venus de Medici, and the Dying Gladiator were at the
Louvre, and when the First Napoleon visited the Senate in
full state.   "He was sitting in one corner of his carriage,
covered and almost hidden from sight by an enormous robe

of ermine, and his face overshadowed by a tremendous
plume of feathers that descended from a velvet hat." We
watch Sir Humphry as Ampère and others bring to him the
first specimens of iodine, and he makes experiments with
his travelling apparatus on the dark lustrous crystals and
their violet vapour ; we seem, too, to be present with the
great English chemist and his scholar as they burn diamonds
at Florence by means of the Grand Duke's gigantic lens,
and prove that the invisible result is carbonic acid ; or as
they study the springs of inflammable gas at Pietra Mala,
and the molten minerals of Vesuvius. The whole, too, is
interspersed with bits of fun, and this culminates at the
Roman Carnival, where he evidently thoroughly enjoyed the
follies of the Corso, the pelting with sugar-plums, and the
masked balls, to the last of which he went in a nightgown
and nightcap, with a lady who knew all his acquaintances ;
and between the two they puzzled their friends mightily.

This year and a half may be considered as the time of
Faraday's education ; it was the period of his life that best
corresponds with the collegiate course of other men who
have attained high distinction in the world of thought. But
his University was Europe ; his professors the master whom
he served, and those illustrious men to whom the renown of
Davy introduced the travellers. It made him personally
known, also, to foreign *savants*, at a time when there was
little intercourse between Great Britain and the Continent ;
and thus he was associated with the French Academy of
Sciences while still young, his works found a welcome all
over Europe, and some of the best representatives of foreign
science became his most intimate friends.

In May 1815, his engagement at the Royal Institution was renewed, with a somewhat higher position and increased salary, which was again raised in the following year to 100*l.* per annum. The handwriting in the Laboratory Note-book changes in September 1815, from the large running letters of Brande to the small neat characters of Faraday, his first entry having reference to an analysis of " Dutch turf ash," and then soon occur investigations into the nature of sub-stances bearing what must have been to him the mysterious names of Paligenetic tincture, and *Baphe eugenes chruson.* It is to be hoped that the constituents of this golden dye agreed together better than the Greek words of its name.

We can imagine the young philosopher taking a deeper interest in the researches on flame which his master was then carrying out, and in the gradual perfection of the safety-lamp that was to bid defiance to the explosive gases of the mine ; this at least is certain, that Davy, in the preface to his cele-brated paper on the subject, expresses himself " indebted to Mr. Michael Faraday for much able assistance," and that the youthful investigator carefully preserved the manuscript given him to copy.

Part of his duty, in fact, was to copy such papers ; and as Sir Humphry had a habit of destroying them, he begged leave to keep the originals, and in that way collected two large volumes of precious manuscripts.

But there came a change. Hitherto he had been absorb-ing ; now he was to emit. The knowledge which had been a source of delight to himself must now overflow as a bless-ing to others : and this in two ways. His first lecture was given at the City Philosophical Society on January 17, 1816, and in the same year his first paper was published in the *Quarterly Journal of Science.* The lecture was on the general properties of matter; the paper was an analysis of some native caustic lime from Tuscany. Neither was important in itself, but each resembled those little streams which travellers are taken to look at because they are the sources of mighty rivers, for Faraday became the prince of experimental lecturers, and his long series of published researches have won for him the highest niche in the temple of science.

When he began to investigate for himself, it could not have been easy to separate his own work from that which he was expected to do for his master. Hence no small danger of misunderstandings and jealousies; and some of these ugly attendants on rising fame did actually throw their black shadows over the intercourse between the older and the younger man of genius. In these earlier years, however, all appears to have been bright; and the following letter, written from Rome in October 1818, will give a good idea of the assistant's miscellaneous duties, and of the pleasant feelings of Davy towards him. It may be added that in another letter he is requested to send some dozens of "flies with pale bodies" to Florence, for Sir Humphry loved fly-fishing as well as philosophy.

"To Mr. Faraday.

"I received the note you were so good as to address to me at Venice; and by a letter from Mr. Hatchett I find that you have found the parallax of Mr. West's Sirius, and that, as I expected, he is mistaken.

C

"If when you write to me you will give the 3 per cents. and *long annuities*, it will be enough.

"I will thank you to put the enclosed letters into the post, except those for Messrs. Morland and Messrs. Drummond, which perhaps you will be good enough to deliver.

"Mr. Hatchett's letter contained praises of you which were very gratifying to me; and pray believe me there is no one more interested in your success and welfare than your sincere well-wisher and friend,

"H. Davy.

"Rome."

It must not be supposed, however, that he had any astronomical duties, for the parallax he had found was not that of the Dog-star, but of a reputed new metal, Sirium, which was resolved in Faraday's hands into iron, nickel, and sulphur. But the impostor was not to be put down so easily, for he turned up again under the *alias* of Vestium; but again he was unable to escape the vigilant eye of the young detective, for one known substance after another was removed from it; and then, says Faraday, "my Vestium entirely disappeared."

His occupations during this period were multifarious enough. We must picture him to ourselves as a young-looking man of about thirty years of age, well made, and neat in his dress, his cheerfulness of disposition often breaking out in a short crispy laugh, but thoughtful enough when something important is to be done. He has to prepare the apparatus for Brande's lectures, and when the hour has arrived he stands on the right of the Professor, and helps him

to produce the strange transformations of the chemical art.
And conjurers, indeed, the two appear in the eyes of the
youth on the left, who waits upon them, then the "laboratory assistant," now the well-known author, Mr. William
Bollaert, from whom I have learnt many details of this period.
When not engaged with the lectures, Faraday is manufacturing rare chemicals, or performing commercial analyses,
or giving scientific evidence on trials. One of these was a
famous one, arising from the Imperial Insurance Company
resisting the claim of Severn and King, sugar-bakers; and
in it appeared all the chemists of the day, like knights in
the lists, on opposite sides, ready to break a lance with
each other.

All his spare time Faraday was occupied with original
work. Chlorine had a fascination for him, though the yellow
choking gas would get out into the room, and he investi-
gated its combinations with carbon, squeezed it into a
liquid, and applied it successfully as a disinfectant when
fatal fever broke out in the Millbank Penitentiary. Iodine
too, another of Davy's elements, was made to join itself to
carbon and hydrogen; and naphthaline was tormented with
strong mineral acids. Long, too, he tried to harden steel
and prevent its rusting, by alloying it with small quantities
of platinum and the rarer metals; the boy blew the bellows
till the crucibles melted, but a few ordinary razors seem to
have been the best results. Far more successful was he in
repeating and extending some experiments of Ampère on
the mutual action of magnets and electric currents; and
when, after months of work and many ingenious contrivances, the wire began to move round the magnet, and the

magnet round the wire, he himself danced about the revolving metals, his face beaming with joy—a joy not unmixed
with thankful pride—as he exclaimed, "There they go!
there they go! we have succeeded at last." After this discovery he thought himself entitled to a treat, and proposed
to his attendant a visit to the theatre. "Which shall it
be?" "Oh, let it be Astley's, to see the horses." So to
Astley's they went; but at the pit entrance there was
a crush; a big fellow pressed roughly upon the lad, and
Faraday, who could stand no injustice, ordered him to
behave himself, and showed fight in defence of his young
companion.

The rising philosopher indulged, too, in other recreations.
He had a wonderful velocipede, a progenitor of the modern
bicycle, which often took him of an early morning to Hamp-
stead Hill. There was also his flute; and a small party for
the practice of vocal music once a week at a friend's house.
He sang bass correctly, both as to time and tune.

And though the City Philosophical Society was no more,
the ardent group of students of nature who used to meet
there were not wholly dispersed. They seem to have carried on their system of mutual improvement, and to have
read the current scientific journals at Mr. Nicol's house till
he married, and then alternately at those of Mr. R. H.
Solly, Mr. Ainger, and Mr. Hennel, of Apothecaries' Hall,
who came to a tragical end through an explosion of fulmi-
nating silver. Several of them, including Mr. Cornelius
Varley, joined the Society of Arts, which at that time had
committees of various sciences, and was very democratic in
its management; and, finding that by pulling together they

had great influence, they constituted themselves a "caucus,"
adopting the American word, and meeting in private.
Magrath was looked upon as a " chair-maker," and Faraday
in subsequent years held the office of Chairman of the
Committee of Chemistry, and occasionally he presided at
the large meetings of the Society.

During this time (1823) the Athenæum Club was started,
not in the present Grecian palace in Pall Mall, but in a private house in Waterloo Place.   Its members were the aristocracy of science, literature, and art, and they made Faraday
their honorary secretary ; but after a year he transferred the
office to his friend Magrath, who held it for a long period.

Among the various sects into which Christendom is
divided, few are less known than the Sandemanians. About
a century and a half ago, when there was little light in the
Presbyterian Church of Scotland, a pious minister of the
name of John Glas began to preach that the Church should
be governed only by the teaching of Christ and His apostles,
that its connection with the State was an error, and that we
ought to believe and to practise no more and no less than
what we find from the New Testament that the primitive
Church believed and practised.   These principles, which
sound very familiar in these days, procured for their asserter
much obloquy and a deposition by the Church Courts, in
consequence of which several separate congregations were
formed in different parts of Great Britain, especially by
Robert Sandeman, the son-in-law of Mr. Glas, and from
him they received their common appellation.   In early
days they taught a simpler view of faith than was generally
held at that time ; it was with them a simple assent of

the understanding, but produced by the Spirit of God, and its virtue depended not on anything mystical in the operation itself, but on the grandeur and beauty of the things believed. Now, however, there is little to distinguish them in doctrine from other adherents of the Puritan theology, though they certainly concede a greater deference to their elders, and attach more importance to the Lord's Supper than is usual among the Puritan Churches. Their form of worship, too, resembles that of the Presbyterians; but they hold that each congregation should have a plurality of elders, pastors, or bishops, who are unpaid men; that on every "first day of the week" they are bound to assemble, not only for prayers and preaching, but also for "breaking of bread," and putting together their weekly offerings; that the love-feast and kiss of charity should continue to be practised; that "blood and things strangled" are still forbidden as food; and that a disciple of Christ should not charge interest on loans, or lay up wealth for the unknown future, but rather consider all he possesses as at the service of his poorer brethren, and be ready to perform to them such offices of kindness as in the early Church were expressed by washing one another's feet.

But what gives the remarkable character to the adherents of this sect is their perfect isolation from all Christian fellowship outside their own community, and from all external religious influence. They have never made missionary efforts to win men from the world, and have long ceased to draw to themselves members from other Churches; so they have rarely the advantage of fresh blood, or fresh views of the meaning of Scripture. They constantly inter-

marry, and are expected to " bear one another's burthens ; "
so the Church has assumed the additional character of a
large intertwined family and of a mutual benefit society. This
rigid separation from the world, extending now through three
or four generations, has produced a remarkable elevation of
moral tone and refinement of manner ; and it is said that no
one unacquainted with the inner circle can conceive of the
brotherly affection that reigns there, or the extent to which
hospitality and material help is given without any ostenta-
tion, and received without any loss of self-respect. The
body is rendered still more seclusive by demanding, not
merely unity of spirit among its members, but unanimity of
opinion in every Church transaction. In order to secure
this, any dissentient who persists in his opinion after repeated argument is rejected : the same is also the consequence
of neglect of Church duties, as well as of any grave
moral offence : and in such a community excommunication
is a serious social ban, and though a penitent may be received back once, he can never return a second time.

It was in the midst of this little community that Faraday
received his earliest religious impressions, and among them
he found his ecclesiastical home till the day of his entrance
into the Church above.

Among the elders of the Sandemanian Church in London
was Mr. Barnard, a silversmith, of Paternoster Row. The
young philosopher became a visitor at his house, and though
he had previously written,

> " What is't that comes in false deceitful guise,
> Making dull fools of those that 'fore were wise ?
>   'Tis Love,"

he altered his opinion in the presence of the citizen's third daughter, Sarah, and wrote to her what was certainly not the letter of a fool :—

" You know me as well or better than I do myself. You know my former prejudices and my present thoughts—you know my weaknesses, my vanity, my whole mind ; you have converted me from one erroneous way, let me hope you will attempt to correct what others are wrong. . . . . Again and again I attempt to say what I feel, but I cannot. Let me, however, claim not to be the selfish being that wishes to bend your affections for his own sake only. In whatever way I can best minister to your happiness, either by assiduity or by absence, it shall be done. Do not injure me by withdrawing your friendship, or punish me for aiming to be more than a friend by making me less ; and if you cannot grant me more, leave me what I possess,—but hear me."

The lady hesitated, and went to Margate. There he followed her, and they proceeded together to Dover and Shakspeare's Cliff, and he returned to London full of happiness and hope. He loved her with all the ardour of his nature, and in due course, on June 12, 1821, they were married. The bridegroom desired that there should be no bustle or noise at the wedding, and that the day should not be specially distinguished ; but he calls it himself " an event which more than any other contributed to his happiness and healthful state of mind." As years rolled on the affection between husband and wife became only deeper and deeper; his bearing towards her proved it, and his letters frequently testify to it. Doubtless at any time between their marriage

and his final illness he might have written to her as he did
from Birmingham, at the time of the British Association :—
" After all, there is no pleasure like the tranquil pleasures of
home, and here—even here—the moment I leave the table,
I wish I were with you IN QUIET. Oh ! what happiness is
ours ! My runs into the world in this way only serve to
make me esteem that happiness the more."

He took his bride home to Albemarle Street, and there
they spent their wedded life ; but until Mr. Barnard's death
it was their custom to go every Saturday to the house of the
worthy silversmith, and spend Sunday with him, returning
home usually in the evening of that day. His own father
died while he was at Riebau's, but his mother, a grand-
looking woman, lived long afterwards, supported by her
son, whom she occasionally visited at the Institution, and
of whose growing reputation she was not a little proud.

With a mind calmed and strengthened by this beautiful
domestic life, he continued with greater and greater enthusiasm to ask questions of Nature, and to interpret her re-
plies to his fellow-men. Just before his marriage he had
been appointed at the Royal Institution superintendent of
the house and laboratory, and in February 1825, after a
change in the management of the Institution, he was placed
as director in a position of greater responsibility and influence. One of his first acts in this capacity was to invite
the members to a scientific evening in the laboratory ; this
took place three or four times in 1825, and in the following
years these gatherings were held every week from Feb. 3
to June 9 ; and though the labour devolved very much
upon Faraday, other philosophers sometimes brought for-

ward discoveries or useful inventions. Thus commenced those Friday evening meetings which have done so much to popularize the high achievements of science. Faraday's note-books are still preserved, containing the minutes of the committee-meetings every Thursday afternoon, the Duke of Somerset chairman, and he secretary; also the record of the Friday evenings themselves, who lectured, and on what subject, and what was exhibited in the library, till June 1840, when other arrangements were probably made.

The year 1827 was otherwise fruitful in lectures: in the spring, a course of twelve on chemical manipulation at the London Institution; after Easter, his first course at Albemarle Street, six lectures on chemical philosophy (he had helped Professor Brande in 1824);[1] and at Christmas, his desire to convey knowledge, and his love to children, found expression in a course of six lectures to the boys and girls home for their holidays. These were a great success; indeed, he himself says they "were just what they ought to have been, both in matter and manner,—but it would not answer to give an extended course in the same spirit." He continued these juvenile lectures during nineteen years. The notes for courses of lectures were written in school copy-books, and sometimes he appends a general remark about the course, not always so favourable as the one given above. Thus he writes, "The eight lectures on the operations of the laboratory, April 1828, were not to my mind."

---

[1] Sir Roderick Murchison used to tell how he was attending Brande's lectures, when one day, the Professor being absent, his assistant took his place, and lectured with so much ease that he won the complete approval of the audience. This, he said, was Faraday's first lecture at the Royal Institution.

Of the course of twelve in the spring of 1827, he says he
" found matter enough in the notes for at least seventeen."

Up to 1833 Faraday was bringing the forces of nature in
subjection to man on a salary of only 100*l.* per annum,
with house, coals, and candles, as the funds of the Institution would not at that time afford more ; but among the
sedate *habitués* of the place was a tall, jovial gentleman, who
lounged to the lectures in his old-fashioned blue coat and
brass buttons, grey smalls, and white stockings, who was a
munificent friend in need. This was John Fuller, a member
of Parliament. He founded a Professorship of Chemistry,
with an endowment that brings in nearly 100*l.* a year, and
gave the first appointment to Faraday for life. When the
Institution became richer, his income was increased ; and
when, on account of the infirmities of age, he could no
longer investigate, lecture, or keep accounts, the managers
insisted on his still retaining in name his official connection
with the place, with his salary and his residence there. Nor
indeed could they well have acted otherwise; for though
the Royal Institution afforded in the first instance a congenial soil for the budding powers of Faraday, his growth
soon became its strength ; and eventually the blooming of
his genius, and the fruit it bore, were the ornament and
glory of the Institution.

It will be asked, Was this 100*l.* or 200*l.* per annum the
sole income of Faraday ? No ; in early days he did commercial analyses, and other professional work, which paid
far better than pure science. In 1830 his gains from this
source amounted to 1,000*l.*, and in 1831 to considerably
more ; they might easily have been increased, but at that

time he made one of his most remarkable discoveries—the evolution of electricity from magnetism,[1]—and there seemed to lie open before him the solution of the problem how to make one force exhibit at will the phenomena of magnetism or of common or voltaic electricity. And then he had to face another problem—his own mental force might be turned either to the acquisition of a fortune, or to the following up of those great discoveries; it would not do both : which should he relinquish? The choice was deliberately made : Nature revealed to him more and more of her secrets, but his professional gains sank in 1832 to 155*l.* 9*s.*, and during no subsequent year did they amount even to that.

Still his work was not entirely confined to his favourite studies. In a letter to Lord Auckland, long afterwards, he says :—" I have given up, for the last ten years or more, all professional occupation, and voluntarily resigned a large income that I might pursue in some degree my own objects of research. But in doing this I have always, as a good subject, held myself ready to assist the Government if still in my power, *not for pay;* for, except in one instance (and then only for the sake of the person joined with me), I refused to take it. I have had the honour and pleasure of applications, and that very recently, from the Admiralty, the Ordnance, the Home Office, the Woods and Forests, and other departments, all of which I have replied to, and will reply to as long as strength is left me." He had declined the Professorship of Chemistry at the London University—now University College,—but in 1829 he ac-

---

[1] The laboratory note-book shows that at this very time he was making a long series of commercial analyses of saltpetre for Mr. Brande.

cepted a lectureship at the Royal Academy, Woolwich, and held it for about twenty years. In 1836 he became scientific adviser to the Trinity House, and his letter to the Deputy Master also shows his feelings in reference to such employment :—" You have left the title and the sum in pencil. These I look at mainly as regards the character of the appointment ; you will believe me to be sincere in this, when you remember my indifference to your proposition as a matter of interest, though *not as a matter of kindness.* In consequence of the goodwill and confidence of all around me, I can at any moment convert my time into money, but I do not require more of the latter than is sufficient for necessary purposes. The sum, therefore, of 200*l.* is quite enough in itself, but not if it is to be the indicator of the character of the appointment; but I think you do not view it so, and that you and I understand each other in that respect ; and your letter confirms me in that opinion. The position which I presume you would wish me to hold is analogous to that of a standing counsel." For nearly thirty years Faraday continued to report on all scientific suggestions and inventions connected with lighthouses or buoys, not for personal gain or renown, but for the public good. His position was never above that of a " standing counsel." In his own words : " I do not know the exact relation of the Board of Trade and the Trinity House to each other; I am simply an adviser upon philosophical questions, and am put into action only when called upon."

In regard to the lectureship at Woolwich, Mr. Abel, his successor, writes thus :—" Faraday appears to have enjoyed his weekly trips to Woolwich, which he continued for so

many years, as a source of relaxation.   He was in the habit
of going to Woolwich in the afternoon or evening preceding his lecture at the Military Academy, then preparing
at once for his experiments, and afterwards generally taking
a country ramble.   The lecture was delivered early the following morning.   No man was so respected, admired, and
beloved as a teacher at the Military Academy in former
days as Faraday.   Many are the little incidents which have
been communicated to me by his pupils illustrative of his
charms as a lecturer, and of his kindly feelings for the
youths to whom he endeavoured to impart a taste for, if not
a knowledge of, science.   But for some not ill-meant, though
scarcely judicious, proposal to dictate modifications in his
course of instruction, Faraday would probably have continued
for some years longer to lecture at Woolwich.   In May
1852, soon after I had been appointed his successor,
Faraday wrote to me requesting the return of some tubes
of condensed gases which he left at the Academy.   This
letter ends thus :—' I hope you feel yourself happy and
comfortable in your arrangements at the Academy, and have
cause to be pleased with the change.   I was ever very kindly
received there, and that portion of regret which one must
ever feel in concluding a long engagement would be in some
degree lessened with me by hearing that you had reason to
be satisfied with your duties and their acceptance.—Ever
very truly yours, M. FARADAY.' "

For year after year the life of Faraday afforded no adven-
ture and little variety, only an ever-growing skill in his
favourite pursuits, higher and higher success, and ever-
widening fame.   But simple as were his mind and his

habits, no one picture can present him as the complete man ; we must try to make sketches from various points of view, and leave it to the reader's imagination to combine them.

Let us watch him on an ordinary day. After eight hours' sleep, he rises in time to breakfast at eight o'clock, goes round the Institution to see that all is in order, and descends into the laboratory, puts on a large white apron full of holes, and is busy among his pieces of apparatus. The faithful Anderson, an old soldier, who always did exactly what he was told, and nothing more,[1] is waiting upon him ; and as thought flashes after thought through his eager—perhaps impatient—brain, he twists his wires into new shapes, and rearranges his magnets and batteries. Then some conclusion is arrived at which lights up his face with a gleam of satisfaction, but the next minute a doubt comes across that expressive brow—may the results not be due to something else yet imperfectly conceived ?—and a new experiment must be devised to answer that. In the meantime one of his little nieces has been left in his charge. She sits as quiet as a mouse with her needlework ; but now and then he gives her a nod, or a kind word, and throwing a little piece of potas-

---

[1] The following anecdote has been sent me on the authority of Mr. Benjamin Abbott :—" Sergeant Anderson was engaged to attend to the furnaces in Mr. Faraday's researches on optical glass in 1828, and was chosen simply because of the habits of strict obedience his military training had given him. His duty was to keep the furnaces always at the same heat, and the water in the ashpit always at the same level. In the evening he was released, but one night Faraday forgot to tell Anderson he could go home, and early next morning he found his faithful servant still stoking the glowing furnace, as he had been doing all night long."

sium on to a basin of water for her amusement, he shows
her the metal bursting into purple flame, floating about in
fiery eddies, and the crack of the fused globule of potash at
the end.   Presently there is handed to him the card of some
foreign *savant*, who makes his pilgrimage to the famous
Institution and its presiding genius ; he puts down his last
result on a slate, comes upstairs, and, disregarding the inter-
ruption, chats with his visitor with all cordiality and open-
ness.   Then to work again till dinner-time, at half-past two.
In the afternoon he retires to his study with its plain, furniture and the india-rubber tree in the window, and writes a
letter full of affection to some friend, after which he goes off
to the council meeting of one of the learned bodies.   Then
back again to the laboratory, but as evening approaches he
goes upstairs to his wife and niece, and then there is a game
at bagatelle or acting charades ; and afterwards he will read
aloud from Shakspeare or Macaulay till it is time for supper
and the simple family worship which now is not liable to the
interruptions that generally prevent it in the morning.   And
so the day closes.

Or if it be a fine summer evening, he takes a stroll with
his wife and the little girl to the Zoological Gardens,
and looks at all the new arrivals, but especially the
monkeys, laughing at their tricks till the tears run down
his cheeks.

But should it be a Friday evening, Faraday's place is in
the library and theatre of the Institution, to see that all is
right and ready, to say an encouraging word to the lecturer,
and to welcome his friends as they arrive ; then taking his
seat on the front bench near the right hand of the speaker,

he listens with an animated countenance to his story,[1] sometimes bending forwards, and scarcely capable of keeping
his fingers off the apparatus—not at all able if anything
seems to be going wrong; when the discourse is over, a
warm shake of the hand, with "Thank you for a pleasant
hour," and "Good night" to those around him, and upstairs
with his wife and some particularly congenial friends to
supper. On the dining-table is abundance of good fare and
good wine, and around it flows a pleasant stream of lively
and intellectual conversation.

But suppose it is his own night to lecture. The subject has been carefully considered, an outline of his discourse has been written on a sheet of foolscap, with all the
experiments marked and numbered, and during the morning everything has been arranged on the table in such
order that his memory is assisted by it; the audience now
pours in, and soon occupies all the seats, so that late
comers must be content with sitting on the stairs or standing in the gangways, or at the back of the gallery. Faraday
enters, and placing himself in the centre of the horse-shoe
table, perfect master of himself, his apparatus, and his
audience, commences a discourse which few that are present
will ever forget. Here is a picture by Lady Pollock :—" It
was an irresistible eloquence, which compelled attention
and insisted upon sympathy. It waked the young from their
visions, and the old from their dreams. There was a gleaming

---

[1] One evening, when the Rev. A. J. D'Orsey was lecturing "On the
Study of the English Language," he mentioned as a common vulgarism
that of using "don't" in the third person singular, as "He don't pay
his debts." Faraday exclaimed aloud, "That's very wrong."

in his eyes which no painter could copy, and which no poet
could describe.   Their radiance seemed to send a strange
light into the very heart of his congregation; and when
he spoke, it was felt that the stir of his voice and the
fervour of his words could belong only to the owner of
those kindling eyes.   His thought was rapid, and made
itself a way in new phrases—if it found none ready made—
as the mountaineer cuts steps in the most hazardous ascent
with his own axe.   His enthusiasm sometimes carried him
to the point of ecstasy when he expatiated on the beauties
of Nature, and when he lifted the veil from her deep mys-
teries.   His body then took motion from his mind; his
hair streamed out from his head; his hands were full of
nervous action; his light, lithe body seemed to quiver with
its eager life.   His audience took fire with him, and every
face was flushed.   Whatever might be the after-thought or
the after-pursuit, each hearer for the time shared his zeal
and his delight."[1]

Is it possible that he can be happier when lecturing to
the juveniles?   The front rows are filled with the young
people; behind them are ranged older friends and many
of his brother philosophers, and there is old Sir James
South, who is quite deaf, poor man, but has come, as he
says, because he likes to see the happy faces of the children.
How perfect is the attention!   Faraday, with a beaming
countenance, begins with something about a candle or a
kettle that most boys and girls know, then rises to what
they had never thought of before, but which now is as clear
as possible to their understandings.   And with what delight

[1] The *St. Paul's Magazine*, June 1870.

does he watch the performances of Nature in his experiments! One could fancy that he had never seen the experiments before, and that he was about to clap his hands with boyish glee at the unexpected result! Then with serious face the lecturer makes some incidental remark that goes far beyond natural philosophy, and is a lesson for life.

Some will remember one of these occasions which forms the subject of a painting by Mr. Blaikley. Within the circle of the table stands the lecturer, and waiting behind is the trusty Anderson, while the chair is occupied by the Prince Consort, and beside him are the young Prince of Wales and his brother, the present Duke of Edinburgh; while the Rev. John Barlow and Dr. Bence Jones sit on the left of the Princes, Sir James South stands against the door, and Murchison, De la Rue, Mrs. Faraday, and others may be recognized amongst the eager audience.

Let us now suppose that it is a Sunday on which we are watching this prince among the aristocracy of intellect, and we will assume it to be during one of the periods of his eldership, namely between 1840 and 1844, or after 1860. The first period came to a close through his separation both from his office and from the Church itself. The reason of this is said to have been that one Sunday he was absent from the love feast, and, on inquiry being made, it appeared not only that he had been the guest of the Queen, but that he was ready to justify his own conduct in obeying her commands. He, however, continued to worship among his friends, and was after a while restored to the rights of membership, and eventually to the office of elder. In

the morning he and his family group find their way down
to the plain little meeting-house in Paul's Alley, Red-cross
Street, since pulled down to make way for the Metropolitan
Railway. The day's proceedings commence with a prayer
meeting, during which the worshippers gradually drop in
and go to their accustomed seats, Faraday taking his place
on the platform devoted to the elders: then the more
public service begins; one of a metrical but not rhyming
version of the Psalms is sung to a quaint old tune, the Lord's
Prayer and another psalm follow; he rises and reads in a
slow, reverent manner the words of one of the Evangelists,
with a most profound and intelligent appreciation of their
meaning; or he offers an extempore prayer, expressing
perfect trust and submission to God's will, with deep
humility and confession of sin. It may be his turn to
preach. On two sides of a card he has previously sketched
out his sermon with the illustrative texts, but the congregation does not see the card, only a little Bible in his
hand, the pages of which he turns quickly over, as, fresh
from an earnest heart, there flows a discourse full of devout
thought, clothed largely in the language of Scripture. After
a loud simultaneous " Amen " has closed the service, the
Church members withdraw to their common meal, the feast
of charity; and in the afternoon there is another service,
ending by invariable custom with the Lord's Supper. The
family group do not reach home till half-past 5 ; then there
is a quiet evening, part of which is spent by Faraday at his
desk, and they retire to rest at an early hour.

Again on Wednesday evening he is among the little flock.
The service is somewhat freer, for not the officers of the

Church only, but the ordinary members are encouraged to express whatever thoughts occur to them, so as to edify one another. At these times, Faraday, especially when he was not an elder, very often had some word of exhortation, and the warmth of his temperament would make itself felt, for he was known in the small community as an experimental rather than a doctrinal preacher.

The notes of his more formal discourses which I have had the opportunity of seeing, indicate, as might be expected from the tenets of his Church, a large acquaintance with the words of Scripture, but no knowledge of modern exegesis. They appear to have impressed different hearers in different ways. One who heard him frequently and was strongly attached to him, says that his sermons were too parenthetical and rapid in their delivery, with little variety or attractiveness; but another scientific friend, who heard him occasionally, writes: "They struck me as resembling a mosaic work of texts. At first you could hardly understand their juxtaposition and relationship, but as the well-chosen pieces were filled in, by degrees their congruity and fitness became developed, and at last an amazing sense of the power and beauty of the whole filled one's thoughts at the close of the discourse."

Among the latest of his sermons was one that he preached at Dundee about four years before his death. He began by telling his audience that his memory was failing, and he feared he could not quote Scripture with perfect accuracy; and then, as said one of the elders present, "his face shone like the face of an angel," as he poured forth the words of loving exhortation.

When a mind is stretched in the same direction week-day and Sunday, the tension is apt to become too great. With Faraday the first symptom was loss of memory. Then his devoted wife had to hurry him off to the country for rest of brain. Once he had to give up work almost entirely for a twelvemonth. During this time he travelled in Switzerland, and extracts from his diary are given by Bence Jones. His niece, Miss Reid, gives us her recollections of a month spent at Walmer:—" How I rejoiced to be allowed to go there with him ! We went on the outside of the coach, in his favourite seat behind the driver. When we reached Shooter's Hill, he was full of fun about Falstaff and the men in buckram, and not a sight nor a sound of interest escaped his quick eye and ear. At Walmer we had a cottage in a field, and my uncle was delighted because a window looked directly into a blackbird's nest built in a cherry-tree. He would go many times in a day to watch the parent birds feeding their young. I remember, too, how much he was interested in the young lambs, after they were sheared at our door, vainly trying to find their own mothers. The ewes, not knowing their shorn lambs, did not make the customary signal. In those days I was eager to see the sun rise, and my uncle desired me always to call him when I was awake. So, as soon as the glow brightened over Pegwell Bay, I stole downstairs and tapped at his door, and he would rise, and a great treat it was to watch the glorious sight with him. How delightful, too, to be his companion at sunset ! Once I remember well how we watched the fading light from a hill clothed with wild flowers, and how, as twilight stole on, the sounds of bells

from Upper Deal broke upon our ears, and how he watched
till all was grey. At such times he would be well pleased
if we could repeat a few lines descriptive of his feelings."
And then she tells us about their examining the flowers
in the fields by the aid of "Galpin's Botany," and how with
a candle he showed her a spectre on the white mist outside
the window; of reading lessons that ended in laughter, and
of sea-anemones and hermit crabs, with the merriment
caused by their odd movements as they dragged about the
unwieldy shells they tenanted. "But of all things I used
to like to hear him read 'Childe Harold;' and never shall
I forget the way in which he read the description of the
storm on Lake Leman. He took great pleasure in Byron,
and Coleridge's 'Hymn to Mont Blanc' delighted him.
When anything touched his feelings as he read—and it
happened not unfrequently—he would show it not only
in his voice, but by tears in his eyes also."

A few days at Brighton refreshed him for his work. He
was in the habit of running down there before his juvenile
lectures at Christmas, and at Easter he frequently sought
the same sea-breezes.

But it was not always that Faraday could run away from
London when the mental tension became excessive. A
shorter relaxation was procured by his taking up a novel,
such as "Ivanhoe," or "Jane Eyre," or "Monte Christo."
He liked the stirring ones best, "a story with a thread to
it." Or he would go with his wife to see Kean act, or hear
Jenny Lind sing, or perhaps to witness the performance of
some "Wizard of the North."

Now and then he would pay a visit to some scene of

early days. One of his near relatives tells me : " It is said
that Mr. Faraday once went to the shop where his father
had formerly been employed as a blacksmith, and asked to
be allowed to look over the place. When he got to a part
of the premises at which there was an opening into the
lower workshop, he stopped and said : ' I very nearly lost
my life there once. I was playing in the upper room at
pitching halfpence into a pint pot close by this hole, and
having succeeded at a certain distance, I stepped back to try
my fortune further off, forgetting the aperture, and down I
fell ; and if it had not been that my father was working over
an anvil fixed just below, I should have fallen on it, broken
my back, and probably killed myself. As it was, my father's
back just saved mine.' "

Business, as well as pleasure, sometimes took him away
from home. He often joined the British Association, returning usually on Saturday, that he might be among his
own people on the Lord's Day. During the meeting he
would generally accept the hospitality of some friend ; and
it was one of these occasions that gave rise to the following
jeu d'esprit :—

" ' That P will change to F in the British tongue is true
    (Quoth Professor Phillips), though the instances are few : '
An entry in my journal then I ventured thus to parody,
' I this day dined with Fillips, where I hobbed and nobbed with
    Pharaday.'

" OXFORD, *June* 27, 1860."   " T. T.

At the Liverpool meeting, in 1837, he was president of
the Chemical Section, and on two other occasions he was

selected to deliver the evening lecture, but though repeatedly
pressed to undertake the presidency of the whole body, he
could not be prevailed upon to accept the office.

My first personal intercourse with him, of any extent, was
at the Ipswich meeting, in 1851. I watched him with all
the interest of an admiring disciple, and there is deeply en-
graven on my memory the vivacity of his conversation, the
eagerness with which he entered into some mathematico-
chemical speculations of Dumas, and the playfulness with
which, when we were dining together, he cut boomerangs out
of card, and shot them across the table at his friends.

Professional engagements also took him not unfrequently
into the country. Some of these will be described in the
later sections, that treat of his mode of working and its
valuable results.

To comprehend a man's life it is necessary to know not
merely what he does, but also what he purposely leaves
undone. There is a limit to the work that can be got out
of a human body or a human brain, and he is a wise man
who wastes no energy on pursuits for which he is not fitted ;
and he is still wiser who, from among the things that he
can do well, chooses and resolutely follows the best.

Faraday took no part in any of the political or social
movements of his time. To politics indeed he seems to
have been really indifferent. It was during the intensely
interesting period of 1814-15 that he was on the Continent
with Davy, but he alludes to the taking of Paris by the
allied troops simply because of its bearing on the movements
of the travellers, and on March 7, 1815, he made this
remarkable entry in his journal: " I heard for news that

Bonaparte was again at liberty. Being no politician, I did not trouble myself much about it, though I suppose it will have a strong influence on the affairs of Europe." In later days he seems to have awaked to sufficient interest to read the debates, and to show a Conservative tendency; he became a special constable in 1848, and was disposed generally to support " the powers that be,"—though that involved some perplexity at a change of Government.

It is more singular that a man of his benevolent spirit should never have taken a prominent part in any philanthropic movement. During the latter half of his life, he, as a rule, avoided serving on committees even for scientific objects, and was reluctant to hold office in the learned societies with which he was connected. I believe, however, that this arose not from want of interest, but from a conviction that he was ill-suited by natural temperament for joining in discussions on subjects that roused the passions of men, or for calmly weighing the different causes of action, and deciding which was the most judicious. It is remarkable how little even of his scientific work was done in conjunction with others. Neither did he spend time in rural occupations, or in literary or artistic pursuits. Beasts and birds and flowers he looked at, but it was for recreation, not for study. Music he was fond of, and occasionally he visited the Opera, but he did not allow sweet sounds to charm him away from his work. He stuck closely to his fireside, his laboratory, his lecture table, and his Church. He lived where he worked, so that he had only to go downstairs to put to the test of experiment any fresh thought that flitted across his brain. He almost invariably declined dinner-parties, except at Lady Davy's, and at Mr. and Mrs. Masquerier's at Brighton, towards whom he felt under an obligation on account of former kindnesses. If he went to a *soirée*, he usually stayed but a short time; and even when away from home he generally refused private hospitality. Thus he was able to give almost undivided attention to the chief pursuit of his life.

His residence in so accessible a part of London did, however, expose him to the constant invasion of callers, and his own good nature often rendered fruitless the efforts that were considerately made to restrict these within reasonable limits. Of course he suffered from the curious and the inconsiderate of the human species; and then there were those pertinacious bores, the dabblers in science. "One morning a young man called on him, and with an air of great importance confided to him the result of some original researches (so he deemed them) in electrical philosophy. 'And pray,' asked the Professor, taking down a volume of Rees' Cyclopædia, 'did you consult this or any elementary work to learn whether your discovery had been anticipated?' The young man replied in the negative. 'Then why do you come to waste my time about well-known facts, that were published forty years ago?' 'Sir,' said the visitor, 'I thought I had better bring the matter to head-quarters immediately.' 'All very well for you, but not so well for head-quarters,' replied the Professor, sharply, and set him down to read the article."

"A grave, elderly gentleman once waited upon him to submit to his notice 'a new law of physics.' The visitor requested that a jug of water and a tumbler might be brought, and then producing a cork, 'You will be pleased to observe,' said he, 'how persistently this cork clings to the side of the glass when the vessel is half filled.' 'Just so,' replied the Professor. 'But now,' resumed this great discoverer, 'mark what happens when I fill the tumbler to the brim. There! you see the cork flies to the centre—positively repelled by the sides!' 'Precisely so,' replied the amused electrician, with the air of a man who felt perfectly at home with the phenomenon, and indeed regarded it quite as an old friend. The visitor was evidently disconcerted. 'Pray how long have you known this?' he ventured to ask Faraday. 'Oh, ever since I was a boy,' was the rejoinder. Crestfallen—his discovery demolished in a moment—the poor gentleman was retiring with many apologies, when the Professor, sincerely concerned at his disappointment, comforted him by suggesting that possibly he might some day alight upon something really new."[1]

But there were other visitors who were right welcome to a portion of his time. One day it might be a young man, whom a few kind words and a little attention on the part of the great philosopher would send forward on the journey of life with new energy and hopes. Another day it might be some intellectual chieftain, who could meet the prince of experimenters on equal terms. But these are hardly to be regarded as interruptions;—rather as part of his chosen work.

Here is one instance in the words of Mr. Robert Mallet. ". . . . I was, in the years that followed, never in London without paying him a visit, and on one of those times I

[1] *British Quarterly Review*, April 1868.

ventured to ask him (if not too much engaged) to let me see
where he and Davy had worked together. With the most
simple graciousness he brought me through the whole of the
Royal Institution, Albemarle Street. Brande's furnaces,
Davy's battery, the place in the laboratory where he told
me he had first observed the liquefaction of chlorine, are all
vividly before me—but nothing so clear or vivid as our
conversation over a specimen of green (crown) glass, partially devitrified in floating opaque white spheres of radiating
crystals : he touched luminously on the obscure relation of
the vitreous and crystalloid states, and on the probable
nature of the nuclei of the white spheres. My next visit to
Faraday that I recollect was not long after my paper 'On the
Dynamics of Earthquakes' had appeared in the Transactions
of the Royal Irish Academy. He almost at once referred
to it in terms of praise that seemed to me so far beyond
my due, that even now I recall the very humble way I
felt, as the thought of Faraday's own transcendent merits
rushed across my mind. I ventured to ask him, had the
paper engaged his attention sufficiently that I might ask
him—did he consider my explanation of the before supposed *vorticose* shock sufficient? To my amazement he at
once recited *nearly word for word* the paragraph in which
I took some pains to put my views into a demonstrative
shape, and ended with, 'It is as plain and certain as a
proposition of Euclid !' And yet the subject was one
pretty wide away from his own objects of study."

Often, too, if some interesting fact was exhibited to him,
he would send to his brother *savants* some such note as
this :—

"ROYAL INSTITUTION, 4*th May*, 1852.

'MY DEAR WHEATSTONE,

"Dr. Dubois-Raymond will be making his experiments *here* next Thursday, the 6th, from and after 11 o'clock. I wish to let you know, that you may if you like join the select few.

"Ever truly yours,

"M FARADAY."

It was indeed his wont to share with others the delight to a new discovery. Thus Sir Henry Holland tells me that he used frequently to run to his house in Brook Street with some piece of scientific news. One of these visits was after reading Bunsen and Kirchhoff's paper on Spectrum Analysis; and he did not stop short with merely telling the tale of the special rays of light shot forth by each metallic vapour, as the following letter will show. It is addressed to the present Baroness Burdett Coutts.

"ROYAL INSTITUTION, *Friday*, 17*th May.*

"DEAR MISS COUTTS,

"To-morrow, at 4 o'clock, immediately after Max Müller's lecture, I shall show Sir Henry Holland an apparatus which has arrived from Munich to manifest the phenomena of light which have recently been made known to us by Bunsen and Kirchhoff. Mr. Barlow will be here, and he suggests that you would like to know of the occasion. If you are inclined to see how philosophers work and live, and so are inclined to climb our narrow stairs (for I must show

the experiments in my room), we shall be most happy to
see you. The experiments will not be beautiful except to
the intelligent.

" Ever your faithful Servant,
"M. FARADAY."

Sometimes, too, the exhibition of a scientific fact would
take him away from home. Thus, when her Majesty and
the Prince Consort once paid a private visit to the Polytechnic, Mr. Pepper arranged a surprise for the Royal party,
by getting Faraday in a quiet room to explain the Ruhmkorff's coil—the latest development of his own inductive
currents. This he did with his usual vivacity and enthusiasm, and the interview is said to have gratified the philosopher as well as the Queen.

He could not, however, escape the inroads made upon
his time by correspondence. People would write and ask
him questions. Once a solitary prisoner wrote to tell him,
" It is indeed in studying the great discoveries which science
is indebted to you for, that I render my captivity less sad,
and make time flow with rapidity,"—and then he proceeds
to ask, " *What is the most simple* combination to give to a
voltaic battery, in order to produce a spark capable of setting
fire to powder under water, or under ground? Up to the
present I have only seen employed to that purpose piles of
thirty to forty pairs constructed on Dr. Wollaston's principles. They are very large and inconvenient for field
service. Could not the same effect be produced by two spiral
pairs only? and if so, what can be their smallest dimension? "
And who was the prisoner who thus speculated on the

applications of science to war? It was no other than Prince Louis Napoleon, then immured in the fortress of Ham, and now the ex-Emperor of the French. At another time he wrote asking for his advice in the manufacture of an alloy which should be about as soft as lead, but not so fusible,—a question which also had evident bearing upon the art of war; and offering at the same time to pay the cost of any experiments that might be necessary.

Often, too, the correspondents of Faraday thought that they were doing him a kindness. He says somewhere: " The number of suggestions, hints for discovery, and pro- positions of various kinds, offered to me very freely and with perfect goodwill and simplicity on the part of the proposers, for my exclusive investigation and final honour, is remarkably great, and it is no less remarkable that but for one exception—that of Mr. Jenkin—they have all been worthless . . . . I have, I think, universally found that the man whose mind was by nature or self-education fitted to make good and worthy suggestions, was also the man both able and willing to work them out."

Both the askers of questions and the givers of advice expected answers—and the answers came. Most of Fara- day's letters, indeed, are of a purely business character: sometimes they are very laconic, as the note in which he announced to Dr. Paris one of his principal discoveries :—

" DEAR SIR,
            " The *oil* you noticed yesterday turns out to be liquid chlorine.   "Yours faithfully,
                                    " M. FARADAY."

But in other letters, as may be expected, there is found the enthusiasm of his ardent nature, or the glow of his genial spirit.   An instance or two may suffice.

" ROYAL INSTITUTION, 24th March, 1843.

" DEAR SIR,

"I have received and at once looked at your paper.   Many thanks for so good a contribution to the beloved science.   What glorious steps electricity has taken in the days within our remembrance, and what hopes are held out for the future !   The great difficulty is to remove the mists which dim the dawn of a subject, and I cannot but consider your paper as doing very much that way for a most important part of natural knowledge.

" I am, my dear Sir,

" Most truly yours,

" M. FARADAY.

" J. P. JOULE, ESQ."

" ROYAL INSTITUTION, 15th Oct. 1853.

" MY DEAR MISS MOORE,

" The summer is going away, and I never (but for one day) had any hopes of profiting by your kind offer of the roof of your house in Clarges Street.   What a feeble summer it has been as regards sunlight !   I have made a good many preliminary experiments at home, but they do not encourage me in the direction towards which I was looking.   All is misty and dull, both the physical and the mental prospect.   But I have ever found that the experimental philosopher has great need of patience, that he may not be downcast by interposing obstacles, and perseverance, that

E

he may either overcome them, or open out a new path to the
bourn he desires to reach. So perhaps next summer I may
think of your housetop again. Many thanks for your kind
letter and all your kindnesses uswards. My wife had your
note yesterday, and I enjoyed the violets, which for a time
I appropriated.

"With kindest remembrances and thoughts to all with you
and her at Hastings,

"I am, my dear Friend,

"Very faithfully yours,

"M. FARADAY."

The following is written to Mr. Frank Barnard, then an
Art student in Paris :—

"ROYAL INSTITUTION, 9th Nov., 1852.

"MY DEAR NEPHEW,

"Though I am not a letter-writer and shall not
profess to send you any news, yet I intend to waste your
time with one sheet of paper : first to thank you for your
letter to me, and then to thank you for what I hear of your
letters to others. You were very kind to take the trouble of
executing my commissions, when I know your heart was
bent upon the entrance to your studies. Your account of
M. Arago was most interesting to me, though I should have
been glad if in the matter of health you could have made it
better. He has a wonderful mind and spirit. And so you
are hard at work, and somewhat embarrassed by your position : but no man can do just as he likes, and in many
things he has to give way, and may do so honourably,
provided he preserve his self-respect. Never, my dear

Frank, lose that, whatever may be the alternative. Let no one tempt you to it; for nothing can be expedient that is not right; and though some of your companions may tease you at first, they will respect you for your consistency in the end; and if they pretend not to do so, it is of no consequence. However, I trust the hardest part of your probation is over, for the earliest is usually the hardest; and that you know how to take all things quietly. Happily for you, there is nothing in your pursuit which need embarrass you in Paris. I think you never cared for home politics, so that those of another country are not likely to occupy your attention, and a stranger can be but a very poor judge of a new people and their requisites.

"I think all your family are pretty well, but I know you will hear all the news from your appointed correspondent Jane, and, as I said, I am unable to chronicle anything. Still, I am always very glad to hear how you are going on, and have a sight of all that I may see of the correspondence.

"Ever, my dear Frank,

"Your affectionate Uncle,

"M. FARADAY."

His scientific researches were very numerous. The Royal Society Catalogue gives under the name of Faraday a list of 158 papers, published in various scientific magazines or learned Transactions. Many of these communications are doubtless short, but a short philosophical paper often represents a large amount of brain work; a score of them are the substance of his Friday evening discourses; while others are lengthy treatises, the records of long and careful investi-

gations; and the list includes the thirty series of his " Experimental Researches in Electricity."   These extended
over a period of twenty-seven years, and were afterwards
reprinted from the " Philosophical Transactions," and form
three goodly volumes, with 3,430 numbered paragraphs—one
of the most marvellous monuments of intellectual work, one
of the rarest treasure-houses of newly-discovered knowledge,
with which the world has ever been enriched.   Faraday
never published but one book in the common acceptation
of the term—it was on " Chemical Manipulation,"—but
there appeared another large volume of reprinted papers ;
and three of his courses of lectures were also published as
separate small books, though not by himself. It is very tempt-
ing to linger among these 158 papers; but this is not intended
as a scientific biography, and those readers who wish to
make themselves better acquainted with his work will find an
admirable summary of it in Professor Tyndall's " Faraday
as a Man of Science." In Sections IV. and V., however, I
have endeavoured to give an idea of his manner of working
and of the practical benefits that have flowed to mankind
from some of his discoveries.

As these papers appeared his fame grew wider and wider.
When a comparatively young man he was naturally desirous
of appending the mystic letters " F.R.S." to his name, and
he was balloted into the Royal Society in January 1824, not
without strong opposition from his master, Sir Humphry
Davy, then president. He paid the fees, and never sought
another distinction of the kind. But they were showered down
upon him. The Philosophical Society of Cambridge had
already acknowledged his merits, and the learned Academies

of Paris and Florence had enrolled him amongst their corresponding members. Heidelberg and St. Petersburg, Philadelphia and Boston, Copenhagen, Berlin, and Palermo, quickly followed ; and as the fame of his researches spread, very many other learned societies in Europe and America, as well as at home, brought to him the tribute of their honorary membership.[1] He thrice received the degree of Doctor, Oxford making him a D.C.L., Prague a Ph.D., and Cambridge an LL.D., besides which he was instituted a Chevalier of the Prussian Order of Merit, a Commander of the Legion of Honour, and a Knight Commander of the Order of St. Maurice and St. Lazarus. Among the medals which he received were each of those at the disposal of the Royal Society—indeed the Copley medal was given him twice—and the Grande Médaille d'Honneur at the time of the French Exhibition. Altogether it appears he was decorated with ninety-five titles and marks of merit,[2] including the blue ribbon of science, for in 1844 he was chosen one of the eight foreign associates of the French Academy.

Though he had never passed through a university career, he was made a member of the Senate of the University of London, which he regarded as one of his chief honours ; and he showed his appreciation of the importance of the office by a diligent attendance to its duties.

As the recognized prince of investigators, it is no wonder

[1] See Appendix.

[2] No wonder the celebrated electrician P. Riess, of Berlin, once addressed a long letter to him as "Professor Michael Faraday, Member of all Academies of Science, London."

that, on the resignation of Lord Wrottesley, an attempt was made to induce him to become President of the Royal Society. A deputation waited on him and urged the unanimous wish of the Council and of scientific men. Faraday begged for time to consider. Tyndall gives us an insight into the reasons that led him to decline. He tells us: "On the following morning I went up to his room, and said, on entering, that I had come to him with some anxiety of mind. He demanded its cause, and I responded, 'Lest you should have decided against the wishes of the deputation that waited on you yesterday.' 'You would not urge me to undertake this responsibility,' he said. 'I not only urge you,' was my reply, 'but I consider it your bounden duty to accept it.' He spoke of the labour that it would involve; urged that it was not in his nature to take things easy; and that if he became president, he would surely have to stir many new questions, and agitate for some changes. I said that in such cases he would find himself supported by the youth and strength of the Royal Society. This, however, did not seem to satisfy him. Mrs. Faraday came into the room, and he appealed to her. Her decision was adverse, and I deprecated her decision. 'Tyndall,' he said at length, 'I must remain plain Michael Faraday to the last; and let me now tell you, that if I accepted the honour which the Royal Society desires to confer upon me, I would not answer for the integrity of my intellect for a single year.'"

In 1835 Sir Robert Peel desired to confer pensions as honourable distinctions on Faraday and some other eminent men. Lord Melbourne, who succeeded him as Prime

Minister, in making the offer at a private interview, gave utterance to some hasty expressions that appeared to the man of science to reflect on the honour of his profession, and led to his declining the money. The King, William IV., was struck with the unusual nature of the proceeding, and kept repeating the story of Faraday's refusal; and about a month afterwards the Premier, dining with Dr. (now Sir Henry) Holland, begged him to convey a letter to the Professor and to press on him the acceptance of the pension. The letter was couched in such honourable and conciliatory terms, that Faraday's personal objection could no longer apply, and he expressed his willingness to receive this mark of national approval. A version of the matter that found its way into the public prints caused fresh annoyance, and nearly produced a final refusal, but through the kind offices of friends who had interested themselves throughout in the matter, a friendly feeling was again arrived at, and the pension of £300 a year was granted and accepted.

In 1858 the Queen offered him a house at Hampton Court. It was a pretty little place, situated in the well-known Green in front of the Palace; and in that quiet retreat Faraday spent a large portion of his remaining years.

In October 1861 he wrote a letter to the managers of the Royal Institution, resigning part of his duties, in which he reviewed his connection with them. "I entered the Royal Institution in March 1813, nearly forty-nine years ago, and, with exception of a comparatively short period, during which I was abroad on the Continent with Sir H. Davy, have been with you ever since. During that time I

have been most happy in your kindness, and in the fostering care which the Royal Institution has bestowed upon me.
Thank God, first, for all His gifts. I have next to thank
you and your predecessors for the unswerving encouragement and support which you have given me during that
period. My life has been a happy one, and all I desired.
During its progress I have tried to make a fitting return
for it to the Royal Institution, and through it to science.
But the progress of years (now amounting in number to three-
score and ten) having brought forth first the period of development, and then that of maturity, have ultimately produced
for me that of gentle decay. This has taken place in such
a manner as to make the evening of life a blessing; for
whilst increasing physical weakness occurs, a full share of
health free from pain is granted with it; and whilst
memory and certain other faculties of the mind diminish,
my good spirits and cheerfulness do not diminish with
them."

When he could no longer discharge effectually his duties
at the Trinity House, the Corporation quietly made their
arrangements for transferring them, and, with the concur-
rence of the Board of Trade, determined that his salary of
200*l.* per annum should continue as long as he lived. Sir
Frederick Arrow called upon him at Albemarle Street, and
explained how the matter stood, but he found it hard to
persuade the Professor that there was no injustice in his
continuing to receive the money; then, taking hold of Sir
Frederick by one hand and Dr. Tyndall by the other,
Faraday, with swimming eyes, passed over his office to
his successor.

Gradually but surely the end approached. The loss of
memory was followed by other symptoms of declining power.
The fastenings of his earthly tabernacle were removed one
by one, and he looked forward to "the house not made
with hands, eternal in the heavens." This was no new
anticipation. Calling on the friend who had long directed
with him the affairs of the Institution, but who was then
half paralysed, he had said, " Barlow, you and I are waiting ;
that is what we have to do now ; and we must try to do it
patiently." He had written to his niece, Mrs. Deacon : " I
cannot think that death has to the Christian anything in it
that should make it a rare, or other than a constant, thought ;
out of the view of death comes the view of the life beyond
the grave, as out of the view of sin (that true and real view
which the Holy Spirit alone can give to a man) comes the
glorious hope. . . . . . My worldly faculties are slipping
away day by day. Happy is it for all of us that the true
good lies not in them. As they ebb, may they leave us as
little children trusting in the Father of Mercies, and accepting His unspeakable gift." And when the dark shadow was
creeping over him, he wrote to the Comte de Paris : " I bow
before Him who is Lord of all, and hope to be kept waiting
patiently for His time and mode of releasing me according
to His Divine Word, and the great and precious promises
whereby His people are made partakers of the Divine nature."
His niece, Miss Jane Barnard, who tended him with most
devoted care, thus wrote from Hampton Court on the 27th
June :—" The kind feelings shown on every side towards my
dear uncle, and the ready offers of help, are most soothing.
I am thankful to say that we are going on very quietly ; he

keeps his bed and sleeps much, and we think that the paralysis gains on him, but between whiles he speaks most pleasant words, showing his comfort and trust in the finished work of our Lord. The other day he repeated some verses of the 46th Psalm, and yesterday a great part of the 23rd. We can only trust that it may be given us to say truly, 'Thy will be done;' indeed, the belief that all things work together for good to them that believe, is an anchor of hope, sure and steadfast, to the soul. We are surrounded by most kind and affectionate friends, and it is indeed touching to see what warm feelings my dear uncle has raised on all sides."

When his faculties were fading fast, he would sit long at the western window, watching the glories of the sunset; and one day when his wife drew his attention to a beautiful rainbow that then spanned the sky, he looked beyond the falling shower and the many-coloured arch, and observed, "He hath set his testimony in the heavens." On August 25, 1867, quietly, almost imperceptibly, came the release. There was a philosopher less on earth, and a saint more in heaven.

The funeral, at his own request, was of the simplest character. His remains were conveyed to Highgate Cemetery by his relations, and deposited in the grave, according to the practice of his Church, in perfect silence. Few of his scientific friends were in London that bright summer-time, but Professor Graham and one or two others came out from the shrubbery, and joining the group of family mourners, took their last look at the coffin.

But when this sun had set below our earthly horizon,

there seemed to spring up in the minds of men a great desire to catch some of the rays of the fading brightness and reflect them to posterity. A " Faraday Memorial" was soon talked of, and the work is now in the sculptor's hands; the Chemical Society has founded a " Faraday Lectureship;" one of the new streets in Paris has been called " Rue Faraday;" biographical sketches have appeared in many of the British and Continental journals; successive books have told the story of his life and work; and in a thousand hearts there is embalmed the memory of this Christian gentleman and philosopher.

SECTION II.

STUDY OF HIS CHARACTER.

In the previous section we have traced the leading events of a life which was quietly and uniformly successful. We have watched the passage of the errand-boy into the philosopher, and we have seen how at first he begged for the meanest place in a scientific workshop, and at last declined the highest honour which British Science was capable of granting. His success did not lie in the amassing of money—he deliberately turned aside from the path of proffered wealth; nor did it lie in the attainment of social position and titles;—he did not care for the weight of these. But if success consists in a life full of agreeable occupation, with the knowledge that its labours are adding to the happiness and wealth of the world, leading on to an old age full of honour, and the prospect of a blissful immortality,—then the highest success crowned the life of Faraday.

How did he obtain it? Not by inheritance, and not by the force of circumstances. The wealth or the reputation of fathers is often an invaluable starting-point for sons : a liberal education and the contact of superior minds in

early youth is often a mighty help to the young aspirant :
the favour of powerful friends will often place on a vantage
ground the struggler in the battle of life. But Faraday had
none of these. Accidental circumstances sometimes push
a man forward, or give him a special advantage over his
fellows ; but Faraday had to make his circumstances, and
to seize the small favours that fortune sometimes threw in
his way. The secret of his success lay in the qualities of
his mind.

It is only fair, however, to remark that he started with no
disadvantages. There was no stain in the family history :
he had no dead weight to carry, of a disgraced name, or of
bad health, or deficient faculties, or hereditary tendencies to
vice. It must be acknowledged, too, that he was endowed
with a naturally clear understanding and an unusual power
of looking below the surface of things.

The first element of success that we meet with in his
biography is the faithfulness with which he did his work.
This led the bookseller to take his poor errand-boy as an
apprentice ; and this enabled his father to write, when he
was 18 : " Michael is bookbinder and stationer, and is very
active at learning his business. He has been most part of
four years of his time out of seven. He has a very good master
and mistress, and likes his place well. He had a hard time
for some while at first going; but, as the old saying goes,
he has rather got the head above water, as there is two other
boys under him." This faithful industry marked also his
relations with Davy and Brande, and the whole of his subsequent life, and at last, when he found that he could no
longer discharge his duties, it made him repeatedly press his

resignation on the managers of the Royal Institution, and beg to be relieved of his eldership in the Church.

His love of study, and hunger after knowledge, led him to the particular career which he pursued, and that power of imagination, which reveals itself in his early letters, grew and grew, till it gave him such a familiarity with the unseen forces of nature as has never been vouchsafed to any other mortal.

As a source of success there stands out also his enthusiasm. A new fact seemed to charge him with an energy that gleamed from his eyes and quivered through his limbs, and, as by induction, charged for the time those in his presence with the same vigour of interest. Plücker, of Bonn, was showing him one day in the laboratory at Albemarle Street his experiments on the action of a magnet on the electric discharge in vacuum tubes. Faraday danced round them ; and as he saw the moving arches of light, he cried, "Oh! to live always in it!" Mr. James Heywood once met him in the thick of a tremendous storm at Eastbourne, rubbing his hands with delight because he had been fortunate enough to see the lightning strike the church tower.

This enthusiasm led him to throw all his heart into his work. Nor was the energy spasmodic, or wasted on unworthy objects, for, in the words of Bence Jones, his was "a lifelong lasting strife to seek and say that which he thought was true, and to do that which he thought was kind."

Indeed, his perseverance in a noble strife was another of the grand elements in his success. His tenacity of purpose

showed itself equally in little and in great things. Arrang-
ing some apparatus one day with a philosophical instrument
maker, he let fall on the floor a small piece of glass : he
made several ineffectual attempts to pick it up. "Never
mind," said his companion, "it is not worth the trouble."
"Well, but, Murray, I don't like to be beaten by something
that I have once tried to do."

The same principle is apparent in that long series of
electrical researches, where for a quarter of a century he
marched steadily along that path of discovery into which he
had been lured by the genius of Davy. And so, whatever
course was set before him, he ran with patience towards the
goal, not diverted by the thousand objects of interest which
he passed by, nor stopping to pick up the golden apples
that were flung before his feet.

This tremendous faculty of work was relieved by a
wonderful playfulness. This rarely appears in his writings,
but was very frequent in his social intercourse. It was a
simple-hearted joyousness, the effervescence of a spirit at
peace with God and man. It not seldom, however, assumed
the form of good-natured banter or a practical joke.
Indications of this playfulness have already been given, and
I have tried to put upon paper some instances that occur
to my own recollection, but the fun depended so much
upon his manner, that it loses its aroma when separated
from himself.

However, I will try one story. I was spending a night at
an hotel at Ramsgate when on lighthouse business. Early in
the morning there came a knock at the bed-room door, but,
as I happened to be performing my ablutions, I cried,

"Who's there?" "Guess." I went over the names of my brother commissioners, but heard only "No, no," till, not thinking of any other friend likely to hunt me up in that place, I left off guessing; and on opening the door I saw Faraday enjoying with a laugh my inability to recognize his voice through a deal board.

A student of the late Professor Daniell tells me that he remembers Faraday often coming into the lecture-room at King's College just when the Professor had finished and was explaining matters more fully to any of his pupils who chose to come down to the table. One day the subject discoursed on and illustrated had been sulphuretted hydrogen, and a little of the gas had escaped into the room, as it perversely will do. When Faraday entered he put on a look of astonishment, as though he had never smelt such a thing before, and in a comical manner said, "Ah! a savoury lecture, Daniell!" On another occasion there was a little ammonia left in a jar over mercury. He pressed Daniell to tell him what it was, and when the Professor had put his head down to see more clearly, he whiffed some of the pungent gas into his face.

Occasionally this humour was turned to good account, as when, one Friday evening before the lecture, he told the audience that he had been requested by the managers to mention two cases of infringement of rule. The first related to the red cord which marks off the members' seats. "The second case I take to be a hypothetical one, namely, that of a gentleman wearing his hat in the drawing-room." This produced a laugh, which the Professor joined in, bowed, and retired.

This faithful discharge of duty, this almost intuitive insight into natural phenomena, and this persevering enthusiasm in the pursuit of truth, might alone have secured a great position in the scientific world, but they alone could never have won for him that large inheritance of respect and love. His contemporaries might have gazed upon him with an interest and admiration akin to that with which he watched a thunderstorm; but who feels his affections drawn out towards a mere intellectual Jupiter? We must look deeper into his character to understand this. There is a law well recognized in the science of light and heat, that a body can absorb only the same sort of rays which it is capable of emitting. Just so is it in the moral world. The respect and love of his generation were given to Faraday because his own nature was full of love and respect for others.

Each of these qualities—his respect for and love to others, or, more generally, his reverence and kindliness—deserves careful examination.

Throughout his life, Michael Faraday appeared as though standing in a reverential attitude towards Nature, Man, and God.

Towards Nature, for he regarded the universe as a vast congeries of facts which would not bend to human theories. Speaking of his own early life, he says: "I was a very lively imaginative person, and could believe in the 'Arabian Nights' as easily as in the 'Encyclopædia;' but facts were important to me, and saved me. I could trust a fact, and always cross-examined an assertion." He was indeed a true disciple of that philosophy which says, "Man, who is the

F

servant and interpreter of Nature, can act and understand no
farther than he has, either in operation or in contemplation,
observed of the method and order of Nature." [1]   And verily
Nature admitted her servant into her secret chambers, and
showed him marvels to interpret to his fellow-men more
wonderful and beautiful than the phantasmagoria of Eastern
romance.

His reverence towards Man showed itself in the respect
he uniformly paid to others and to himself.   Thoroughly
genuine and simple-hearted himself, he was wont to credit
his fellow-men with high motives and good reasons.   This
was rather uncomfortable when one was conscious of no
such merit, and I at least have felt ashamed in his presence
of the poor commonplace grounds of my words and actions.
To be in his company was in fact a moral tonic.   As he
had learned the difficult art of honouring all men, he was
not likely to run after those whom the world counted great.
"We must get Garibaldi to come some Friday evening,"
said a member of the Institution during the visit of the
Italian hero to London.   " Well, if Garibaldi thinks he can
learn anything from us, we shall be happy to see him," was
Faraday's reply.   This nobility of regard not only preserved
him from envying the success of other explorers in the
same field, but led him heartily to rejoice with them in their
discoveries.

Dumas gives us a picture of Foucault showing Faraday
some of his admirable experiments, and of the two men
looking at one another with eyes moistened, but full of
bright expression, as they stood hand in hand, silently

[1] Bacon's "Novum Organum," i. i.

thankful—the one for the pleasure he had experienced, the other for the honour that had been done him. He also tells how, on another occasion, he breakfasted at Albemarle Street, and during the meal Mr. Faraday made some eulogistic remarks upon Davy, which were coldly received by his guest. After breakfast, he was taken downstairs to the ante-room of the lecture theatre, when Faraday, walking up to the portrait of his old master, exclaimed, "Wasn't he a great man!" then turning round to the window next the entrance door, he added, "It was there that he spoke to me for the first time." The Frenchman bowed. They descended the stairs again to the laboratory. Faraday pulled out an old note-book, and turning over its pages showed where Davy had entered the means by which the first globule of potassium was produced, and had drawn a line round the description, with the words, "Capital experiment." The French chemist owned himself vanquished, and tells the tale in honour of him who remembered the greatness and forgot the littlenesses of his teacher.

And the respect he showed to others he required to be shown to himself. It is difficult to imagine anyone taking liberties with him, and it was only in early life that there were small-minded creatures who would treat him not according to what he was, but according to the position from which he had risen. His servants and workpeople were always attentive to the smallest expression of his wish. Still, he did not "go through life with his elbows out." He once wrote to Matteucci: "I see that that moves you which would move me most, viz. the imputation of a want of good faith; and I cordially sympathize with anyone who is so charged unjustly. Such cases have seemed to me almost the only ones for which it is worth while entering into controversy. I have felt myself not unfrequently misunderstood, often misrepresented, sometimes passed by, as in the cases of specific inductive capacity, magneto-electric currents, definite electrolytic action, &c. &c.; but it is only in the cases where moral turpitude has been implied, that I have felt called upon to enter on the subject in reply." Yet, where he felt that his honour was impugned, none could be more sensitive or more resolute.

This desire to clear himself, combined with his delicate regard for the feelings of others, struck me forcibly in the following incident. At Mr. Barlow's one Friday evening after the discourse, two or three other chemists and myself were commenting unfavourably on a public act of Faraday, when suddenly he appeared beside us. I did not hesitate to tell him my opinion. He gave me a short answer, and joined others of the company. A few days afterwards he found me in the laboratory preparing for a lecture, and, without referring directly to what I had said, he gave me a full history of the transaction in such a way as to show that he could not have acted otherwise, and at the same time to render any apology on my part unnecessary.

Intimately connected with his respect for Man as well as reverence for truth, was the flash of his indignation against any injustice, and his hot anger against any whom he discovered to be pretenders. When, for instance, he had convinced himself that the reputed facts of table-turning and spiritualism were false, his severe denunciation of the whole thing followed as a matter of course.

Thus, too, a story is told of his once taking the side of the injured in a street quarrel by the pump in Savile Row. One evening also at my house, a young man who has since acquired a scientific renown was showing specimens of some new compounds he had made. A well-known chemist objected that, after all, they were mere products of the laboratory : but Faraday came to the help of the young experimenter, and contended that they were chemical substances worthy of attention, just as much as though they occurred in nature.

His reverence for God was shown not merely by that homage which every religious man must pay to his Creator and Redeemer, but by the enfolding of the words of Scripture and similar expressions in such a robe of sacredness, that he rarely allowed them to pass his lips or flow from his pen, unless he was convinced of the full sympathy of the person with whom he was holding intercourse.

This characteristic reverence was united to an equally characteristic kindliness. This word does not exactly express the quality intended ; but unselfishness is negative, goodness is too general, love is commonly used with special applications ; kindness, friendship, geniality, and benevolence are only single aspects of the quality. Let the reader add these terms all together, and the resultant will be about what is meant.[1]

Faraday's love to children was one way in which this kindliness was shown. Having no children of his own, he surrounded himself usually with his nieces : we have already had a glimpse of him heartily entering into their play, and

[1] Bence Jones has used the Greek ἀγάπη.

we are told how a word or two from Uncle would clear away all the trouble from a difficult lesson, that a long sum in arithmetic became a delight when he undertook to explain it, and that when the little girl was naughty and rebellious, he could gently win her round, telling her how he used to feel himself when he was young, and advising her to submit to the reproof she was fighting against.   Nor were his own relatives the only sharers of this kindness.   One friend cherishes among his earliest recollections, that of Faraday making for him a fly-cage and a paper purse, which had a real bright half-crown in it.   When the present Mr. Baden Powell was a little fellow of thirteen, he used to give short lectures on chemistry in his father's house, and the philosopher of Albemarle Street liked to join the family audience, and would listen and applaud the experiments heartily.   When one day my wife and I called on him with our children, he set them playing at hide-and-seek in the lecture theatre, and afterwards amused them upstairs with tuning-forks and resounding glasses.   At a *soirée* at Mr. Justice Grove's, he wanted to see the younger children of the family; so the eldest daughter brought down the little ones in their nightgowns to the foot of the stairs, and Faraday expressed his gratification with "Ah! that's the best thing you have done to-night."   And when his faculties had nearly faded, it is remembered how the stroking of his hand by Mr. Vincent's little daughter quickened him again to bright and loving interest.

It would be easy to multiply illustrations of this kindliness in various relations of life.

Here is one of his own telling, where certainly the effect

produced was not owing to any knowledge of how princely
an intellect underlay the loving spirit. It is from a journal
of his tour in Wales :—

"*Tuesday, July 20th.*—After dinner I set off on a ramble
to Melincourt, a waterfall on the north side of the valley,
and about six miles from our inn. Here I got a little
damsel for my guide who could not speak a word of
English. We, however, talked together all the way to the
fall, though neither knew what the other said. I was
delighted with her burst of pleasure as, on turning a
corner, she first showed me the waterfall. Whilst I was
admiring the scene, my little Welsh damsel was busy running about, even under the stream, gathering strawberries.
On returning from the fall I gave her a shilling that I
might enjoy her pleasure : she curtsied, and I perceived
her delight. She again ran before me back to the village,
but wished to step aside every now and then to pull strawberries. Every bramble she carefully moved out of the
way, and ventured her bare feet to try stony paths, that
she might find the safest for mine. I observed her as she
ran before me, when she met a village companion, open
her hand to show her prize, but without any stoppage,
word, or other motion. When we returned to the village
I bade her good-night, and she bade me farewell, both by
her actions and, I have no doubt, her language too."

In a letter which Mr. Abel, the Director of the Chemical
Department of the War Establishment, has sent me, occur
the following remarks :—

"Early in 1849 I was appointed, partly through the kind
recommendation of Faraday, to instruct the senior cadets

and a class of artillery officers in the Arsenal, in practical chemistry. On the occasion of my first attendance at Woolwich, when, having just reached manhood, I was about to deliver my first lecture as a recognized teacher, I was naturally nervous, and was therefore dismayed when on entering the class-room I perceived Faraday, who, having come to Woolwich, as usual, to prepare for his next morning's lecture at the Military Academy, had been prompted by his kindly feelings to lend me the support of his presence upon my first appearance among his old pupils. In a moment Faraday put me completely at my ease; he greeted me heartily, saying, 'Well, Abel, I have come to see whether I can assist you;' and suiting action to word, he bustled about, persisting in helping me in the arrangement of my lecture-table,—and at the close of my demonstration he followed me from pupil to pupil, aiding each in his first attempt at manipulation, and evidently enjoying most heartily the self-imposed duty of assistant to his young *protégé*."

Another scientific friend, Mr. W. F. Barrett, writes :— " My first interview with Mr. Faraday ten years ago left an impression upon me I can never forget. Young student as I then was, thinking chiefly of present work and little of future prospects, and till then unknown to Mr. Faraday, judge of my feelings when, taking my hand in both of his, he said, 'I congratulate you upon choosing to be a *philosopher :* it is an arduous life, but a noble and a glorious one. Work hard, and work carefully, and you will have success.' The sweet yet serious way he said this made the earnestness of work become a very vivid reality, and led me to doubt whether I had not dared to undertake too lofty a pursuit. After this Mr. Faraday never forgot to remember me in a number of thoughtful and delicate ways. He would ask me upstairs to his room to describe or show him the results of any little investigation I might have made : taking the greatest interest in it all, his pleasure would seem to equal and thus heighten mine, and then he would add words of kind suggestion and encouragement. In the same kindly spirit he has invited me to his house at Hampton Court, or would ask me to join him at supper after the Friday evening's lecture. His kindness is further shown by his giving me a volume of his researches on Chemistry and Physics, writing therein, 'From his friend Michael Faraday.' Those who live alone in London, unknown and uncared-for by any around them, can best appreciate these marks of attention which Mr. Faraday invariably showed, and not only to myself, but equally to my fellow-assistant in the chemical laboratory."

The following instance among many that might be quoted will illustrate his readiness to take trouble on behalf of others. When Dr. Noad was writing his " Manual of Electricity," a doubt crossed his mind as to whether Sir Snow Harris's unit jar gave a true measure of the quantity of electricity thrown into a Leyden jar : he asked Faraday, and his doubt was confirmed. Shortly afterwards he received a letter beginning thus :—

" MY DEAR SIR,

"Whilst looking over my papers on induction, I was reminded of our talk about Harris's unit jar, and

recollected that I had given you a result just the *reverse*
of my old conclusions, and, as I believe, of the truth. I
think the jar *is a true measure,* so long as the circumstances
of position, &c., are not altered ; for its discharge and the
quantity of electricity thus passed on depends on the constant relation of the balls connected with the inner and
outer surface coating to each other, and is independent of
their joint relation to the machine, battery, &c. . . . Perhaps I have not made my view clear, but next time we
meet, remind me of the matter.

" Ever truly yours,

" M. FARADAY."

And just a week afterwards Dr. Noad received a second
letter, surmounted by a neat drawing, and describing at great
length experiments that the Professor had since made in
order to place the matter beyond doubt.

And it was not merely for friends and brother *savants*
that he would take trouble. Old volumes of the *Mechanics'*
Magazine bear testimony to the way in which he was asked
questions by people in all parts of the kingdom, and that
he was accustomed to give painstaking answers to such
letters.

"Do to others as you would wish them to do to you,"
was a precept often on his lips. But I have heard that
he was sometimes charged with transgressing it himself,
inasmuch as he took an amount of trouble for other people
which he would have been very distressed if they had
taken for him.

His charities were very numerous,—not to beggars ; for

them he had the Mendicity Society's tickets,—but to those whose need he knew. The porter of the Royal Institution has shown me, among his treasured memorials, a large number of forms for post-office orders, for sums varying from 5*s.* to 5*l.*, which Faraday was in the habit of sending in that way to different recipients of his thoughtful bounty. Two or three instances have come to my knowledge of his having given more considerable sums of money—say 20*l.*—to persons who he thought would be benefited by them. In some instances the gift was called a loan, but he lent "not expecting again," and entered into the spirit of the injunction, "When thou doest alms, let not thy left hand know what thy right hand doeth."

This principle was in fact stated in one of his letters to a friend: "As a case of distress I shall be very happy to help you as far as my means allow me in such cases; but then I never let my name go to such acts, and very rarely even the initials of my name." His contributions to the general funds of his Church were kept equally secret.

From all these circumstances, therefore, it is impossible to gauge the amount of his charitable gifts; but when it is remembered that for many years his income from different sources must have been 1,000*l.* or 1,200*l.*, that he and Mrs. Faraday lived in a simple manner—comfortably, it is true, but not luxuriously—and that his whole income was disposed of in some way, there can be little doubt that his gifts amounted to several hundred pounds per annum.

But it was not in monetary gifts alone that his kindness to the distressed was shown. Time was spent as freely as money; and an engrossing scientific research would not be allowed to stand in the way of his succouring the sorrowful. Many persons have told me of his self-denying deeds on behalf of those who were ill, and of his encouraging words. He had indeed a heart ever ready to sympathize. Thus, meeting once in the neighbourhood of Hampton Court an old friend who had retired there invalided and was being drawn about in a Bath chair, he is said to have burst into tears.

When eight years ago my wife and my only son were taken away together, and I lay ill of the same fatal disease, he called at my house, and in spite of remonstrances found his way into the infected chamber. He would have taken me by the hand if I had allowed him; and then he sat a while by my bedside, consoling me with his sympathy and cheering me with the Christian hope.

It is no wonder that this kindliness took the hearts of men captive; and this quality was, like mercy, "twice blessed; it blesseth him that gives, and him that takes." The feeling awakened in the minds of others by this kindliness was indeed a source of the purest pleasure to himself; trifling proofs of interest or love could easily move his thankfulness; and he richly enjoyed the appreciation of his scientific labours. This would often break forth in words. Thus in the middle of a letter to A. De la Rive, principally on scientific matters, he writes :—

"Do you remember one hot day, I cannot tell how many years ago, when I was hot and thirsty in Geneva, and you took me to your house in the town and gave me

II.]   *STUDY OF HIS CHARACTER.*   77

a glass of water and raspberry vinegar? That glass of drink is refreshing to me still."

Again : " Tyndall, the sweetest reward of my work is the sympathy and good-will which it has caused to flow in upon me from all quarters of the world."

But to estimate rightly this amiability of character, it must be distinctly remembered that it was not that super-abundance of good-nature which renders some men incapable of holding their own, or rebuking what they know to be wrong. In proof of this his letters to the spiritualists might be quoted ; but the following have not hitherto seen the light. They are addressed to two different parties whose inventions came officially before him.

" You write 'private' on the outside of your official communication, and 'confidential' within. I will take care to respect these instructions as far as falls within my duty ; but I can have nothing private or confidential *as regards the Trinity House*, which is my chief. Whatever opinion I send to them I must accompany with the papers you send me. If therefore you wish anything held back from them, send me another official answer, and I will return you the one I have, marked 'confidential.' Our correspondence is indeed likely to become a little irregular, because your papers have not come to me through the Trinity House. You will feel that I cannot communicate any opinion I may form to you : I am bound to the Trinity House, to whom I must communicate in confidence. I have no objection to your knowing my conclusions ; but the *Trinity House* is the fit judge of the use it may make of them, or the degree of confidence they may think they deserve, or the parties to whom they may choose to communicate them."

By a foot-note it appears that the *private and confidential* communication was returned to the writer, by desire, four days afterwards.

" SIR,

      " I have received your note and read your pamphlet. There is nothing in either which makes it at all desirable to me to see your apparatus, for I have not time to spare to look at a matter two or three times over. In referring to ——, I suppose you refer also to his application to the Trinity House. In that case I shall hear from him *through the Trinity House.* He has, however, certain inquiries (which I have no doubt have gone to him long ago through the Trinity House) to answer before I shall think it necessary to take any further steps in the matter. With these, however, I suppose you have nothing to do.

      " Are you aware that many years ago our Institution was lighted up for months, if not for years together, by oil-gas (or, as you call it, olefiant gas), compressed into cylinders to the extent of thirty atmospheres, and brought to us from a distance? I have no idea that the patent referred to at the bottom of page 9 could stand for an hour in a court of law. I think, too, you are wrong in misapplying the word *olefiant.* It already belongs to a particular gas, and cannot, without confusion, be used as you use it.

                    " I am, Sir,

                          " Your obedient Servant,

                                " M. FARADAY."

" SIR,

"Thanks for your letter. At the close of it you ask me *privately* and confidingly for the encouragement my opinion might give you if *this power* gas-light is fit for lighthouses. I am unable to assent to your request, as my position at the Trinity House requires that I should be able to take up any subject, applications, or documents they may bring before me in a perfectly unbiassed condition of mind.

"I am, Sir,

"Yours very truly,

"M. FARADAY."

The kindliness which shed its genial radiance on every worthy object around, glowed most warmly on the domestic hearth. Little expressions in his writings often reveal it, as when we read in his Swiss journal about Interlaken : "Clout-nail making goes on here rather considerably, and is a very neat and pretty operation to observe. I love a smith's shop, and anything relating to smithery. My father was a smith."

When he was sitting to Noble for his bust, it happened one day that the sculptor, in giving the finishing touches to the marble, made a clattering with his chisels : noticing that his sitter appeared *distrait*, he said that he feared the jingling of the tools had annoyed him, and that he was weary. "No, my dear Mr. Noble," said Faraday, putting his hand on his shoulder, "but the noise reminded me of my father's anvil, and took me back to my boyhood."

This deep affection peeps out constantly in his letters to different members of his family, "bound up together," as

he wrote to his sister-in-law, "in the one hope, and in faith
and love which is in Jesus Christ." But it was towards his
wife that his love glowed most intensely. Yet how can we
properly speak of this sacred relationship, especially as the
mourning widow is still amongst us? It may suffice to catch
the glimpse that is reflected in the following extract from a
letter he wrote to Mrs. Andrew Crosse on the death of her
husband :—

"July 12, 1855.

" . . . Believe that I sympathize with you most
deeply, for I enjoy in my life-partner those things which you
speak of as making you feel your loss so heavily.

"It is the kindly domestic affections, the worthiness, the
mutual aid in sorrow, the mutual joy in happiness that has
existed, which makes the rupture of such a tie as yours so
heavy to bear; and yet you would not wish it otherwise, for
the remembrance of those things brings solace with the
grief. I speak, thinking what my own trouble would be if
I lost my partner; and I try to comfort you in the only way
in which I think I could be comforted.

" M. FARADAY."

There was, as Tyndall has observed, a mixture of chivalry
with this affection. In his book of diplomas he made the
following remarkable entry :—

"25th January, 1847.

"Amongst these records and events, I here insert the
date of one which, as a source of honour and happiness,
far exceeds all the rest. We were *married* on June 12, 1821.

" M. FARADAY."

On the character of Faraday, these two qualities of reverence and kindliness have appeared to me singularly influential. Among the ways in which they manifested themselves was that beautiful combination of firmness and gentleness which has been frequently remarked : intimately associated with them also were his simplicity and truthfulness. These points must have made themselves evident already, but they deserve further illustration.

In his early days, " one Sabbath morning his swift and sober steps were carrying him along the Holborn pavement towards his meeting-house, when some small missile struck him smartly on the hat. He would have thought it an accident and passed on, when a second and a third rap caused him to turn and look just in time to perceive a face hastily withdrawn from a window in the upper story of a closed linendraper's establishment. Roused by the affront, he marched up to the door and rapped. The servant opening it said there was no one at home, but Faraday declared he knew better, and desired to be shown upstairs. Opposition still being made, he pushed on, made his way up through the house, opened the door of an upper room, discovering a party of young drapers' assistants, who at once professed they knew nothing of the motive of this sudden visit. But the hunter had now run his game to earth : he taxed them sharply with their annoyance of wayfarers on the Sabbath, and said that unless an apology were made at once, they should hear from their employer of something much to their disadvantage. An apology was made forthwith." [1]

[1] For this anecdote, and some others in inverted commas, I am indebted to Mr. Frank Barnard.

Long, long after this event, Dr. and Mrs. Faraday, with Dr. Tyndall, were returning one evening from Mr. Gassiot's, on Clapham Common : a dense fog came on, and they did not know where they were. The two gentlemen got out of their vehicle, and walked to a house and knocked. A man appeared, first at a window and afterwards at the door, very angry indeed at the disturbance, and demanding to know their business. Faraday, in his calm, irresistible manner, explained the situation and their object in knocking. The man instantly changed his tone, looked foolish, and muttered something about being in a fright lest his house of business was on fire.

As to simplicity of character : when, in the course of writing this book, I have spoken to his acquaintances about Faraday, the most frequent comment has been in such words as, " Oh ! he was a beautiful character, and so simple-minded." I have tried to ascertain the cause of this simple-mindedness, and I believe it was the consciousness that he was meaning to do right himself, and the belief that others whom he addressed meant to do right too, and so he could just let them see everything that was passing through his mind. And while he knew no reason for concealment, there was no trace of self-conceit about him, nor any pretence at being what he was not. To illustrate this quality is not so easy ; the indications of it, like his humour, were generally too delicate to be transferred to paper ; but perhaps the following letter will do as well as anything else, for there are few philosophers who could have written so naturally about the pleasures of a pantomime and then about his highest hopes :—

" ROYAL INSTITUTION, LONDON, W.
1st *January*, 1857.

" MY DEAR MISS COUTTS,

"You are very kind to think of our pleasure and send us entrance to your box for to-morrow night. We thank you very sincerely, and I mean to enjoy it, for I still have a sympathy with children and all their thoughts and pleasure. Permit me to wish you very sincerely a happy year ; and also to Mrs. Brown. With some of us our greatest happiness will be content mingled with patience ; but there is much happiness in that and the expected end.

"Ever your obliged Servant,

" M. FARADAY." [1]

As to truthfulness : he was not only truthful in the common acceptation of the word, but he did not allow, either in himself or others, hasty conclusions, random assertions, or slippery logic. "At such times he had a way of repeating the suspicious statement very slowly and distinctly, with an air of wondering scrutiny as if it had astonished him. His irony was then irresistible, and always produced a modification of the objectionable phrase."

" An acquaintance rather given to inflict tedious narratives on his friends was descanting to Faraday on the iniquity of some coachman who had set him down the previous night

----

[1] In another letter that Lady Burdett Coutts has kindly sent me, Faraday says : " We had your box once before, I remember, for a pantomime, which is always interesting to me because of the immense concentration of means which it requires." In a third he makes admiring comments on Fechter.

in the middle of a dark and miry road,—'in fact,' said the irksome drawler, 'in a perfect morass; and there I was, as you may imagine, half the night, plunging and struggling to get out of this dreadful morass.' 'More ass you!' rapped out the philosopher at the top of his scale of laughter." This was a rare instance, for it was only when much provoked that he would perpetrate a pun, or depart from the kind courtesy of his habitual talk.

That he was quite ready to give up a statement or view when it was proved by others to be incorrect, is shown by the Preface to the volumes in which are reprinted his " Experimental Researches." " In giving advice," says Miss Reid, " he always went back to first principles, to the true right and wrong of questions, never allowing deviations from the simple straightforward path of duty to be justified by custom or precedent; and he judged himself strictly by the same rule which he laid down for others."

These beauties of character were not marred by serious defects or opposing faults. " He could not be too closely approached. There were no shabby places or ugly corners in his mind." Yet he was very far from being one of those passionless men who resemble a cold statue rather than throbbing flesh and blood. He was no " model of all the virtues," dreadfully uninteresting, and discouraging to those who feel such calm perfection out of their reach. His inner life was a battle, with its wounds as well as its victory. Proud by nature, and quick-tempered, he must have found the curb often necessary; but notwithstanding the rapidity of his actions and thoughts, he knew how to keep a tight rein on that fiery spirit.

I have listened attentively to every remark in disparagement of Faraday's character, but the only serious ones have appeared to me to arise from a misunderstanding of the man, a misunderstanding the more easy because his standard of right and wrong often differed from the notions current around him.   Still, it may be true that his extreme sensitiveness led him sometimes to do scant justice to those who he imagined were treading too closely in his own footsteps ; as, for instance, when Nobili brought out some beautiful experiments on magnetism, just after the short notice of his own discoveries in 1831 which Faraday had sent to M. Hachette, and which was communicated to the Académie des Sciences.   It is true also that, with his great caution and his repugnance to moral evil, he was more disposed to turn away in disgust from an erring companion than to endeavour to reclaim him.   It has also been imputed to him as a fault that he founded no school, and took no young man by the hand as Davy had taken him.   That this was rather his misfortune than his fault, would appear from words he once wrote to Miss Moore : " I have often endeavoured to discover a genius, but have not been very successful, though many cases seemed promising at first." The world would doubtless have been the gainer if he had stamped his own image on the minds of a group of disciples : but a man cannot do everything ; and had Faraday been more of a teacher, he would perhaps have been less of an investigator.

It has been previously remarked that Faraday took little part in social movements, and went little into society, but it must not be supposed that he was by any means unsocial.

It seems probable that his freedom in this matter was somewhat hampered by the principles in which he had been brought up: it is certain that he was restrained by the desire to give all the time and energy he could to scientific research. Yet pleasant stories are told of his occasional appearances at social gatherings. Thus he liked to attend the Royal Academy dinners, and in earlier days he enjoyed the artistic and musical *conversazioni* at Hullmandel's, where Stanfield Turner and Landseer met Garcia and Malibran; and sometimes he joined this pleasant company at supper and charades, at others in their excursions up the river in an eight-oared cutter. Captain Close has described to me how, when the French Lighthouse authorities put up the screw-pile light on the sands near Calais, they invited the Trinity House officers and Faraday to inspect it. A dinner was arranged for them after the inspection, and M. Reynaud proposed the health of the *étranger célèbre*. A young engineer took exception to Faraday being called a stranger—since he had been at St. Cyr he had known the great Englishman well by his works. The Professor replied to the compliment in the language of his hosts, with a few of his happy and kindly remarks. A gentleman high in the diplomatic service, who was present, remarked that Faraday had said many things which were not French, but not a word which ought not to be so.

More unrestricted was Faraday's sympathy with Nature. He felt the poetry of the changing seasons, but there were two aspects of Nature that especially seemed to claim communion with his spirit : he delighted in a thunderstorm, and he experienced a pleasurable sadness as the orange sunset faded into the evening twilight. There are other minds to which both these sensations are familiar, but they seem to have been felt with great intensity by him. No doubt his electrical knowledge added much to his interest in the grand discharges from the thunder-clouds, but it will hardly account for his standing long at a window watching the vivid flashes, a stranger to fear, with his mind full of lofty thoughts, or perhaps of high communings. Sometimes, too, if the storm was at a little distance, he would summon a cab, and, in spite of the pelting rain, drive to the scene of awful beauty.

One clear starry night Captain Close quoted to him the words of Lorenzo in the " Merchant of Venice : "—

> . . . . . " Look, how the floor of heaven
> Is thick inlaid with patines of bright gold ;
> There's not the smallest orb, which thou behold'st,
> But in his motion like an angel sings,
> Still quiring to the young-eyed cherubins :
> Such harmony is in immortal souls ;
> But, whilst this muddy vesture of decay
> Doth grossly close us in, we cannot hear it."

Faraday, who happened not to be familiar with the passage, made his friend repeat it over and over again as he drank in the whole meaning of the poetry, for there is a true sense in which no other mortal had ever opened his ears so fully to the harmony of the universe.

From the plains of mental mediocrity there occasionally rise the mountains of genius, and from the dead level of selfish respectability there stand out now and then the peaks of moral greatness. Neither kind of excellence is so common as we could wish it, and it is a rare coincidence when, as in Socrates, the two meet in the same individual. In Faraday we have a modern instance. There are persons now living who watched this man of strong will and intense feelings raising himself from the lower ranks of society, yet without losing his balance; rather growing in simplicity, disinterestedness, and humility as princes became his correspondents and all the learned bodies of the world vied with each other to do him homage ; still finding his greatest happiness at home, though reigning in the affections of all his fellows,—loving every honest man, however divergent in opinion, and loved most by those who knew him best.

This is the phenomenon. By what theory is it to be accounted for?

The secret did not lie in the nature of his pursuits. This cannot be better shown than in the following incident furnished me by Mrs. Crosse :—" One morning, a few months after we were married, my husband took me to the Royal Institution to call on Mr. and Mrs. Faraday. I had not seen the laboratory there, and the philosopher very kindly took us over the Institution, explaining for my information many objects of interest. His great vivacity and cheeriness of manner surprised me in a man who devoted his life to such abstruse studies, but I have since learnt to know that the highest philosophical nature is often, indeed generally, united with an almost childlike simplicity.

" After viewing the ample appliances for experimental research, and feeling impressed by the scientific atmosphere of the place, I turned and said, ' Mr. Faraday, you must be very happy in your position and with your pursuits, which elevate you entirely out of the meaner aspects and lower aims of common life.'

" He shook his head, and with that wonderful mobility of countenance which was characteristic, his expression of joyousness changed to one of profound sadness, and he replied : 'When I quitted business, and took to science as a career, I thought I had left behind me all the petty meannesses and small jealousies which hinder man in his moral progress ; but I found myself raised into another sphere, only to find poor human nature just the same everywhere—subject to the same weaknesses and the same self-seeking, however exalted the intellect.'

" These were his words as well as I can recollect ; and, looking at that good and great man, I thought I had never seen a countenance which so impressed me with the characteristic of perfect unworldliness. We know how his life proved that this rare qualification was indeed his."

" Childlike simplicity : " " unworldliness." Where was the tree rooted that bore such beautiful blossoms ? Faraday had learnt in the school of Christ to become " a little child," and he loved not the world because the love of the Father was in him.

We have a charming glimpse of this in an extract which Professor Tyndall has given from an old paper in which he wrote his impressions after one of his earliest dinners with the philosopher :—" At two o'clock he came down for me. He, his niece, and myself formed the party. ' I never give dinners,' he said ; ' I don't know how to give dinners ; and I never dine out. But I should not like my friends to attribute this to a wrong cause. I act thus for the sake

of securing time for work, and not through religious motives
as some imagine.' He said grace. I am almost ashamed to
call his prayer a 'saying' of grace. In the language of
Scripture, it might be described as the petition of a son into
whose heart God had sent the Spirit of His Son, and who
with absolute trust asked a blessing from his father. We
dined on roast beef, Yorkshire pudding, and potatoes, drank
sherry, talked of research and its requirements, and of his
habit of keeping himself free from the distractions of society.
He was bright and joyful—boylike, in fact, though he is now
sixty-two. His work excites admiration, but contact with
him warms and elevates the heart. Here, surely, is a strong
man. I love strength, but let me not forget the example of
its union with modesty, tenderness, and sweetness, in the
character of Faraday."

But his religion deserves a closer attention. When an
errand-boy, we find him hurrying the delivery of his newspapers on a Sunday morning so as to get home in time to
make himself neat to go with his parents to chapel : his
letters when abroad indicate the same disposition ; yet he
did not make any formal profession of his faith till a month
after his marriage, when nearly thirty years of age. Of his
spiritual history up to that period little is known, but there
seem to be good grounds for believing that he did not accept
the religion of his fathers without a conscientious inquiry
into its truth. It would be difficult to conceive of his
acting otherwise. But after he joined the Sandemanian
Church, his questionings were probably confined to matters
of practical duty ; and to those who knew him best nothing
could appear stronger than his conviction of the reality of

the things he believed.   In order to understand the life and
character of Faraday, it is necessary to bear in mind not
merely that he was a Christian, but that he was a Sandemanian.   From his earliest years that religious system
stamped its impress deeply on his mind, it surrounded
the blacksmith's son with an atmosphere of unusual purity
and refinement, it developed the unselfishness of his nature,
and in his after career it fenced his life from the worldliness
around, as well as from much that is esteemed as good by
other Christian bodies.   To this small self-contained sect
he clung with warm attachment; he was precluded from
Christian communion or work outside their circle, but his
sympathies at least burst all narrow bounds.   Thus the
Abbé Moigno tells us that at Faraday's request he one day
introduced him to Cardinal Wiseman.   The interview was
very cordial, and his Eminence did not hesitate frankly and
good-naturedly to ask Faraday if, in his deepest conviction,
he believed all the Church of Christ, holy, catholic, and
apostolical, was shut up in the little sect in which he bore
rule.   "Oh no!" was the reply; "but I do believe from
the bottom of my soul that Christ is with us."   There
were other points, too, in his character which reflected the
colouring of the religious school to which he belonged.
Thus, while humility is inseparable from a Christian life,
there is a special phase of that virtue bred of those doctrines
which teach that all our righteousness must be the unmerited
gift of another: these doctrines are strongly insisted upon in
the Sandemanian Church, and this humility was acquired in
an intense degree by its minister.   Again, while all Christians
deplore the terrible amount of folly and sin in the world,

most recognize also a large amount of good, and believe in
progressive improvement; but small communities are apt to
take gloomy views, and so did Faraday, notwithstanding his
personal happiness, and his firm conviction that "there is
One above who worketh in all things, and who governs even
in the midst of that misrule to which the tendencies and
powers of men are so easily perverted."

In writing to Professor Schönbein and a few other kindred
spirits, he would turn naturally enough from scientific to
religious thoughts, and back again to natural philosophy,
but he generally kept these two departments of his mental
activity strangely distinct, though of course it was well
known that the Professor at Albemarle Street was one of
that long line of scientific men, beginning with the *savants*
of the East, who have brought to the Redeemer the gold,
frankincense, and myrrh of their adoration.

But the peculiar features of Faraday's spiritual life are
matters of minor importance: the genuineness of his
religious character is acknowledged by all. We have admired his faithfulness, his amiability of disposition, and his
love of justice and truth: how far these qualities were
natural gifts, like his clearness of intellect, we cannot
precisely tell; but that he exercised constant self-control
without becoming hard, ascended the pathway of fame
without ever losing his balance, and shed around himself
a peculiar halo of love and joyousness, must be attributed in
no small degree to a heart at peace with God, and to the
consciousness of a higher life.

SECTION III.

FRUITS OF HIS EXPERIENCE.

THOSE who loved Faraday would treasure every word that
he wrote, and to them the life and letters which Bence
Jones has given to the world will be inestimable; but
from the multitude who knew him only at a distance, we
can expect no enthusiasm of admiration. Yet all will
readily believe that through the writings of such a genius
there must be scattered nuggets of intellectual gold, even
when he is not treating directly of scientific subjects.
Some of these relate to questions of permanent interest,
and such nuggets it is my aim to separate and lay before
the reader.

When quite a young man he drew the following ideal
portrait :—" The philosopher should be a man willing to
listen to every suggestion, but determined to judge for himself. He should not be biassed by appearances, have no
favourite hypothesis, be of no school, and in doctrine have
no master. He should not be a respecter of persons, but
of things. Truth should be his primary object. If to
these qualities be added industry, he may indeed hope to
walk within the veil of the temple of Nature." This ideal

he must steadily have kept before him, and not unfrequently in after days he gave utterance to similar
thoughts.   Here are two instances, the first from a lecture
thirty years afterwards, the second from a private letter :—
" We may be *sure* of facts, but our interpretation of facts
we should doubt.   He is the wisest philosopher who holds
his theory with some doubt ; who is able to proportion his
judgment and confidence to the value of the evidence set
before him, taking a fact for a fact, and a supposition for a
supposition; as much as possible keeping his mind free
from all source of prejudice, or, where he cannot do this
(as in the case of a theory), remembering that such a source
is there."   The letter is to Mr. Frederick Field, and relates
to a paper on the existence of silver in the water of the
ocean.

"ROYAL INSTITUTION, 21*st October*, 1856.

" MY DEAR SIR,

          " Your paper looks so well, that though I am of
course unable to become security for the facts, I have still
thought it my duty to send it to the Royal Society.
Whether it will appear there or not I cannot say,—no one
can say even for his own papers ; but for my part, I think
that as facts are the foundation of science, however they
may be interpreted, so they are most valuable, and often
more so than the interpretations founded upon them.   I
hope your further researches will confirm those you have
obtained : but I would not be too hasty with them,—rather
wait a while, and make them quite secure.

          " I am, Sir, your obliged Servant,

                    " M. FARADAY."

How pleasant it would have been to peep into his mind, and watch the process by which he was transferred into the very image of his ideal philosopher! He has partially told us the secret in two remarkable lectures, one of which was delivered before the City Philosophical Society when he was only twenty-seven years of age, while the other formed part of a series on Education at Albemarle Street. Copious extracts from the first are given by Dr. Bence Jones; the second was published at the time. In the early lecture, which is " On the Forms of Matter," he points out the advantages and dangers of systematizing, and winds up his remarks with—

" Nothing is more difficult and requires more care than philosophical deduction, nor is there anything more adverse to its accuracy than fixidity of opinion. The man who is certain he is right is almost sure to be wrong, and he has the additional misfortune of inevitably remaining so. All our theories are fixed upon uncertain data, and all of them want alteration and support. Ever since the world began opinion has changed with the progress of things; and it is something more than absurd to suppose that we have a sure claim to perfection, or that we are in possession of the highest stretch of intellect which has or can result from human thought. Why our successors should not displace us in our opinions, as well as in our persons, it is difficult to say; it ever has been so, and from analogy would be supposed to continue so ; and yet, with all this practical evidence of the fallibility of our opinions, all, and none more than philosophers, are ready to assert the real truth of their opinions."

In his discourse entitled "Observations on Mental
Education," like a skilful physician he first determines what
is the great intellectual disease from which the community
suffers—"deficiency of judgment,"—and then he lays down
rules by which each man may attempt his own cure. For
this self-education, "it is necessary that a man examine
himself, and that not carelessly. . . . A first result of this
habit of mind will be an internal conviction of *ignorance in
many things respecting which his neighbours are taught,* and
that his opinions and conclusions on such matters ought to
be advanced with reservation. A mind so disciplined will
be *open to correction upon good grounds in all things,* even in
those it is best acquainted with; and should familiarize
itself with the idea of such being the case. . . . It is right
that we should stand by and act on our principles, but not
right to hold them in obstinate blindness, or retain them
when proved to be erroneous." And then he gives cases
from his own mental history:—"I remember the time when
I believed a spark was produced between voltaic metals as
they approached to contact (and the reasons why it might
be possible yet remain); but others doubted the fact and
denied the proofs, and on re-examination I found reason to
admit their corrections were well founded. Years ago
I believed that electrolites could conduct electricity by a
conduction proper; that has also been denied by many
through long time: though I believed myself right, yet
circumstances have induced me to pay that respect to
criticism as to re-investigate the subject, and I have the
pleasure of thinking that nature confirms my original
conclusions. So, though evidence may appear to pre-

ponderate extremely in favour of a certain decision, it is wise and proper to hear a counter-statement. You can have no idea how often, and how much, under such an impression, I have desired that the marvellous descriptions which have reached me might prove, in some points, correct; and how frequently I have submitted myself to hot fires, to friction with magnets, to the passes of hands, &c., lest I should be shutting out discovery ;—encouraging the strong desire that something might be true, and that I might aid in the development of a new force of nature." He turns then to another evil, and its cure : " The *tendency to deceive ourselves* regarding all we wish for, and the necessity of *resistance to these desires.* . . . . The force of the temptation which urges us to seek for such evidence and appearances as are in favour of our desires, and to disregard those which oppose them, is wonderfully great. In this respect we are all, more or less, active promoters of error." He winds up his remarks upon this subject with the italicized sentence : " I will simply express my strong belief that that point of self-education which consists in teaching the mind to resist its desires and inclinations until they are proved to be right, is the most important of all, not only in things of natural philosophy, but in every department of daily life." He turns then to the necessity of a " habit of forming clear and precise ideas," and of expressing them in " clear and definite language " :—" When the different data required are in our possession, and we have succeeded in forming a clear idea of each, the mind should be instructed to *balance them* one against another, and not suffered carelessly to hasten to a conclusion." " As

H

a result of this wholesome mental condition, we should be able to form a *proportionate judgment;*" that is, one proportionate to the evidence, ranging through all degrees of probability—while he adds: "Frequently the exercise of the judgment ought to end in *absolute reservation.*"

"The education which I advocate," says Faraday, "will require *patience* and *labour of thought* in every exercise tending to improve the judgment. It matters not on what subject a person's mind is occupied, he should engage in it with the conviction that it will require mental labour." "Because the education is internal, it is not the less needful; nor is it more the duty of a man that he should cause his child to be taught, than that he should teach himself. Indolence may tempt him to neglect the self-examination and experience which form his school, and weariness may induce the evasion of the necessary practices; but surely a thought of the prize should suffice to stimulate him to the requisite exertion; and to those who reflect upon the many hours and days devoted by a lover of sweet sounds to gain a moderate facility upon a mere mechanical instrument, it ought to bring a correcting blush of shame if they feel convicted of neglecting the beautiful living instrument wherein play all the powers of the mind."

At the commencement of this discourse the lecturer felt called upon to limit the range of his remarks :—" High as man is placed above the creatures around him, there is a higher and far more exalted position within his view; and the ways are infinite in which he occupies his thoughts about the fears, or hopes, or expectations of a future life. I believe that the truth of that future cannot be brought to

his knowledge by any exertion of his mental powers, however exalted they may be ; that it is made known to him by
other teaching than his own, and is received through simple
belief of the testimony given. Let no one suppose for a
moment that the self-education I am about to commend
in respect of the things of this life extends to any consi-
derations of the hope set before us, as if man by reasoning
could find out God. It would be improper here to enter
upon this subject further than to claim an absolute distinction between religious and ordinary belief. I shall be
reproached with the weakness of refusing to apply those
mental operations which I think good in respect of high
things to the very highest. I am content to bear the
reproach. Yet, even in earthly matters, I believe that 'the
invisible things of Him from the creation of the world are
clearly seen, being understood by the things that are made,
even His eternal power and Godhead;' and I have never
seen anything incompatible between those things of man
which can be known by the spirit of man which is within
him, and those higher things concerning his future which he
cannot know by that spirit." There is of course a certain
truth in this passage ; spiritual discernment is a real thing
possessed by some, and not by others ; yet is there this
absolute distinction between religious and ordinary belief?
Surely there is the same opportunity and the same necessity for careful judgment, and for resistance to prejudice or
preference, when we are weighing the credentials of anything that may come before us purporting to be a revelation
from above ; surely too, if we have satisfied ourselves that
we possess such a revelation, we must seek for the same

clearness of ideas, and must exercise the same patience and labour of thought, if we would understand it aright. That mental discipline which fits us to interpret the works of God cannot but be akin to the intellectual training required for interpreting His word.

Since Faraday thought and wrote, the question of public education has taken a far deeper hold on the feelings and the hopes of the nation, and it is not merely the extent of the instruction, but its nature also, that is discussed. It is held to be no longer right that the minds of our youth should be fed almost exclusively on the dry husks of classic or mediæval knowledge, while the rich banquet of modern discovery remains untasted. Yet it is hard for natural science to gain an honoured place in our venerable scholastic institutions. Faraday, however, had long formed his conclusions on this subject. In one of his Friday evening discourses he says : "The development of the applications of physical science in modern times has become so large and so essential to the well being of man, that it may justly be used as illustrating the true character of pure science as a department of knowledge, and the claims it may have for consideration by Governments, Universities, and all bodies to whom is confided the fostering care and direction of learning. As a branch of learning, men are beginning to recognize the right of science to its own particular place ; for, though flowing in channels utterly different in their course and end from those of literature, it conduces not less, as a means of instruction, to the discipline of the mind, whilst it ministers, more or less, to the wants, comforts, and proper pleasure, both mental and bodily, of every individual of every class in life. Until of late years, the education for, and recognition of it by the bodies which may be considered as governing the general course of all education, have been chiefly directed to it only as it could serve professional services, viz. those which are remunerated by society; but now the fitness of university degrees in science is under consideration, and many are taking a high view of it, as distinguished from literature, and think that it may well be studied for its own sake, *i.e.* as a proper exercise of the human intelligence, able to bring into action and development all the powers of the mind. As a branch of learning, it has (without reference to its applications) become as extensive and varied as literature; and it has this privilege, that it must ever go on increasing."

On the subject of scientific education Faraday was examined by the Public Schools Commission, November 18th, 1862, and his sentiments of course appear in their report. He said to them : "That the natural knowledge which has been given to the world in such abundance during the last fifty years should remain untouched, and that no sufficient attempt should be made to convey it to the young mind growing up and obtaining its first views of those things, is to me a matter so strange that I find it difficult to understand. Though I think I see the opposition breaking away, it is yet a very hard one to overcome. That it ought to be overcome I have not the least doubt in the world." Lord Clarendon asked him : "You think it is now knocking at the door, and there is a prospect of the door being opened?" "Yes," answered Faraday, "and it

will make its way, or we shall stay behind other nations in
our mode of education." He had been led to the conviction that the exclusive attention to literary studies created
a tendency to regard other things as nonsense, or belonging
only to the artisan, and so the mind is positively injured for
the reception of real knowledge. He says: "It is the
highly educated man that we find coming to us again and
again, and asking the most simple question in chemistry or
mechanics; and when we speak of such things as the
conservation of force, the permanency of matter, and the
unchangeability of the laws of nature, they are far from
comprehending them, though they have relation to us in
every action of our lives. Many of these instructed persons
are as far from having the power of judging of these things
as if their minds had never been trained."

He gives his own opinion as to the precise course to be
pursued with great diffidence; but it is evident that he
would begin the education in natural science at a pretty
early age, and in all cases carry it up to a certain point.
One-fifth of a boy's time might be devoted to this purpose
at present, though in less than half a century he thinks
science will deserve and obtain a far larger share. Sup-
posing a boy of eleven years of age and of ordinary intelligence at one of our public schools : " I would teach him,"
he says, " all those things that come before classics in the
programme of the London University,—Mechanics, hydro-
statics, hydraulics, pneumatics, acoustics, and optics. They
are very simple and easily understood when they are looked
at with attention by both man and boy. With a candle, a
lamp, and a lens or two, an intelligent instructor might teach

optics in a very short time ; and so with chemistry. I
should desire all these." Much would depend on the competency and earnestness of the teacher. " Good lectures
might do a great deal. They would at all events remove
the absolute ignorance which sometimes now appears, but
would give a very poor knowledge of natural things."

Perhaps these opinions of one whose lips are now silent
will yet have their weight in the discussion of this question
both in our highest seats of learning and in those educa-
tional parliaments which have been just called into existence in almost every town and district of our country.

From the somewhat disparaging remarks about lectures
quoted above, it must not be supposed that this prince
of lecturers depreciated his office. " Lectures," he said,
" depend entirely for their value upon the manner in which
they are given. It is not the matter, it is not the subject,
so much as the man ; but if he is not competent, and does
not feel that there is a need of competency, to convey his
ideas gently and quietly and simply to the young mind, he
simply throws up obstacles, and will be found using words
which they will not comprehend." These were the words of
his later days, but fortunately he felt " the need of competency " before his own habits were formed, and in four
letters to Abbott we find wonderfully sagacious observations
on the matter, which it would be well for any young lecturer to study. He describes the proper arrangement of a
lecture-room, dwelling on the necessity of good ventilation ;
and then, having considered the fittest subjects for popular
lectures, he turns to the character of the audience, and shows
how that must be studied ; for some expect to be entertained

by the manner of the lecturer as well as his subject, while others care for something which will instruct. He dwells on the superiority of the eye over the ear as a channel of knowledge, and lays down some rules for this kind of instruction, which he of all men subsequently carried out to perfection. "Apparatus is an essential part of every lecture in which it can be introduced. . . . Diagrams and tables, too, are necessary, or at least add in an eminent degree to the illustration and perfection of a lecture. When an experimental lecture is to be delivered, and apparatus is to be exhibited, some kind of order should be observed in the arrangement of them on the lecture table. Every particular part illustrative of the lecture should be in view; no one thing should hide another from the audience, nor should anything stand in the way of or obstruct the lecturer. They should be so placed, too, as to produce a kind of uniformity in appearance. No one part should appear naked and another crowded, unless some particular reason exists and makes it necessary to be so. At the same time the whole should be so arranged as to keep one operation from interfering with another." A good delivery comes in for its share of praise ; "for though to all true philosophers science and nature will have charms innumerable in every dress, yet I am sorry to say that the generality of mankind cannot accompany us one short hour unless the path is strewed with flowers." Then, "a lecturer should appear easy and collected, undaunted and unconcerned, his thoughts about him, and his mind clear and free for the contemplation and description of his subject. His action should not be hasty and violent, but slow, easy, and natural, consisting principally in changes

of the posture of the body, in order to avoid the air of
stiffness or sameness that would otherwise be unavoidable.
His whole behaviour should evince respect for his audience,
and he should in no case forget that he is in their presence."
He allows a lecturer to prepare his discourse in writing, but
not to read it before the audience, and points out how
necessary it is " to raise their interest at the commencement
of the lecture, and by a series of imperceptible gradations,
unnoticed by the company, keep it alive as long as the
subject demands it." This of course forbids breaks in the
argument, or digressions foreign to the main purpose, and
limits the length of the lecture to a period during which the
listeners can pay unwearied attention. He castigates those
speakers who descend so low as " to angle for claps," or who
throw out hints for commendation, and shows that apologies
should be made as seldom as possible. Experiments should
be to the point, clear, and easily understood : " they should
rather approach to simplicity, and explain the established
principles of the subject, than be elaborate and apply to
minute phenomena only. . . . 'Tis well, too, when the lecturer has the ready wit and the presence of mind to turn
any casual circumstance to an illustration of his subject."
But experiments should be explained by a satisfactory theory ;
or if the scientific world is divided in opinion, both sides of
the question ought to be stated with the strongest arguments
for each, that justice may be done and honour satisfied.

Often in later days was his experience in lecturing made
use of for the benefit of others. " If," he once remarked to
a young lecturer, " I said to my audience, ' This stone will
fall to the ground if I open my hand,' I should open my

hand and let it fall.   Take nothing for granted as known;
inform the eye at the same time as you address the ear."   I
remember him once giving me hints on the laying of the
lecture table at the Institution, and telling me that where
possible he was accustomed to arrange the apparatus in such
a way as to suggest the order of the experiments.   An incident told me by Dr. Carpenter will illustrate some of the
foregoing points.   The first time he heard Faraday lecture
at the Royal Institution, the Professor was explaining the
researches of Melloni on radiant heat.   During the discourse
he touched on the refraction and polarization of heat; and
to explain refraction he showed the simple experiment of
fixing some coloured wafers at the bottom of a basin, and
then pouring in water so as to make them apparently rise.
Dr. Carpenter, who had come up from Bristol with grand
ideas of the lectures at Albemarle Street, wondered greatly
at the introduction of so commonplace an experiment.   Of
course there were many other illustrations, and beautiful
ones too.   He went down, however, after the lecture, to
the table, and among the crowd chatting there was an old
gentleman who remarked, " I think the best experiment
to-night was that of the wafers in the basin."

When a young lecturer, Faraday took lessons in elocution
from Mr. Smart, and was at great pains to cure himself of
any defect of pronunciation or manner; for this purpose he
would get a friendly critic to form part of his audience.   On
the fly-leaves of many of the notes of his lectures are
written the reminders—" Stand up "—" Don't talk quick."
Indeed, in early days it was so much a matter of anxiety to
him that everything in his lectures should be as perfect as

possible, that he not only was accustomed to go over everything again and again in his mind, but the difficulty
of satisfying himself used to trouble his dreams. I was
told this, if I am not mistaken, by himself; and it goes
far to explain how his discourses possessed such a fascination.

Some of his feelings in regard to lecturing may be learnt
from the following particulars, for which I am indebted to
Mr. Charles Tomlinson. They relate to a course of lectures
he delivered in 1849 on Statical Electricity. The first
lecture began thus :—" Time moves on, and brings changes
to ourselves as well as to science. I feel that I must soon
resign into the hands of my successors the position which
I now occupy at this table. Indeed, I have long felt how
much rather I would sit among you and be instructed than
stand here and attempt to instruct. I have always felt my
position in this Institution as a very strange one. Coming
after such a man as Davy, and associated with such a man
as Brande, and having had to make a position for myself,
I have always felt myself here in a strange position. You
will wonder why I make these remarks. It is not from any
affectation of modesty that I do so, but I feel that loss of
memory may soon incapacitate me altogether for my duties.
Without, however, troubling you more about myself, let us
proceed to the subject before us, and fall back upon the
beginnings of the wonderful science of electricity. I shall
have to trouble you with very little of them. The facts are
so wonderful that I shall not attempt to explain them." At
the second lecture, " Faraday advanced to the table at three
o'clock, and began to apologize for an obstruction of voice,
which indeed was painfully evident. He said that, 'in an

engagement where the contracting parties were one and many, the one ought not on any slight ground to break his part of the engagement with the many, and therefore, if the audience would excuse his imperfect utterance, he would endeavour—' Murmurs arose : 'Put off the lecture.' Faraday begged to be allowed to go on.   A medical man then rose and said he had given it as his opinion that it would be dangerous to Dr. Faraday to proceed.   Faraday again urged his wish to proceed—said it was giving so much trouble to the ladies, who had sent away their carriages, and perhaps put off other engagements.   On this the whole audience rose as by a single impulse, and a number of persons surrounded Faraday, who now yielded to the general desire to spare him the pain and inconvenience of lecturing."   A fortnight elapsed before he could again make his appearance, but he continued his course later than usual, in order not to deprive his audience of any of the eight lectures he had undertaken to give them.   Prince Albert came to one of these extra lectures.

Faraday's opinion as to the honours due to scientific men from society or from Government, may be gathered from the following extract from a letter written me by his private friend Mr. Blaikley :—" On one occasion, when making some remark in reference to a movement on behalf of science, I inadvertently spoke of the proper honour due to science.   He at once remarked, 'I am not one who considers that science can be honoured.'   I at once saw the point.   His views of the grandeur of truth, when once apprehended, raised it far beyond any honour that man could give it ; but man might honour himself by respecting and acknowledging it."

Professor George Wilson, of Edinburgh, has thus described his first visit to the philosopher : " Faraday was very
kind, showed me his whole laboratory with labours going
on, and talked frankly and kindly ; but to the usual question
of something to do, gave the usual round O answer, and
treated me to a just, but not very cheering animadversion
on the Government of this country, which, unlike that of
every other civilized country, will give no help to scientific
inquiry, and will afford no aid or means of study for
young chemists."

" Take care of your money," was his advice to Mr. Joule,
then another young aspirant to scientific honours, but who
has since rendered the highest service to science, without
leaning on any hopes of Government help or public
support.

But the impressions given in conversation may not be
always correct. Happily there exist his written opinions
on this subject. In a letter addressed to Professor Andrews
of Belfast, and dated 2nd February, 1843, there occurs this
passage :—" As to the particular point of your letter about
which you honour me by asking my advice, I have no
advice to give ; but I have a strong feeling in the matter,
and will tell you what I should do. I have always felt that
there is something degrading in offering rewards for intellectual exertion, and that societies or academies, or even
Kings and Emperors, should mingle in the matter does not
remove the degradation, for the feeling which is hurt is
a point above their condition, and belongs to the respect
which a man owes to himself. With this feeling, I have
never since I was a boy aimed at any such prize ; or even if,

as in your case, they came near me, have allowed them to move me from my course; and I have always contended that such rewards will never move the men who are most worthy of reward. Still, I think rewards and honours *good* if properly distributed, but they should be given for what a man has done, and not offered for what he is to do, or else talent must be considered as a thing marketable and to be bought and sold, and then down falls that high tone of mind which is the best excitement to a man of power, and will make him do more than any commonplace reward. When a man is rewarded for his deserts, he honours those who grant the reward, and they give it not as a moving impulse to him, but to all those who by the reward are led to look to that man for an example."

Eleven years afterwards Faraday expressed similar views, but more fully, in a letter to the late Lord Wrottesley as chairman of the Parliamentary Committee of the British Association :—

"ROYAL INSTITUTION, *March* 10*th*, 1854.
"MY LORD,

"I feel unfit to give a deliberate opinion on the course it might be advisable for the Government to pursue if it were anxious to improve the position of science and its cultivators in our country. My course of life, and the circumstances which make it a happy one for me, are not those of persons who conform to the usages and habits of society. Through the kindness of all, from my Sovereign downwards, I have that which supplies all my need; and in respect of honours, I have, as a scientific man, received from foreign countries and Sovereigns, those which, belonging to very limited and select classes, surpass in my opinion anything that it is in the power of my own to bestow.

"I cannot say that I have not valued such distinctions; on the contrary, I esteem them very highly, but I do not think I have ever worked for or sought after them. Even were such to be now created here, the time is past when these would possess any attraction for me; and you will see therefore how unfit I am, upon the strength of any personal motive or feeling, to judge of what might be influential upon the minds of others. Nevertheless, I will make one or two remarks which have often occurred to my mind.

"Without thinking of the effect it might have upon distinguished men of science, or upon the minds of those who, stimulated to exertion, might become distinguished, I do think that a Government should *for its own sake* honour the men who do honour and service to the country. I refer now to honours only, not to beneficial rewards; of such honours I think there are none. Knighthoods and baronetcies are sometimes conferred with such intentions, but I think them utterly unfit for that purpose. Instead of conferring distinction, they confound the man who is one of twenty, or perhaps fifty, with hundreds of others. They depress rather than exalt him, for they tend to lower the especial distinction of mind to the commonplaces of society. An intelligent country ought to recognize the scientific men among its people as a class. If honours are conferred upon eminence in any class, as that of the law or the army, they should be in this also. The aristocracy of the class should have other distinctions than those of lowly and high-born, rich and poor, yet they should be such as to be worthy of those whom the Sovereign and the country should delight to honour, and, being rendered very desirable and even enviable in the eyes of the aristocracy by birth, should be unattainable except to that of science. Thus much I think the Government and the country ought to do, for their own sake and the good of science, more than for the sake of the men who might be thought worthy of such distinction. The latter have attained to their fit place, whether the community at large recognize it or not.

"But besides that, and as a matter of reward and encouragement to those who have not yet risen to great distinction, I think the Government should, in the very many cases which come before it having a relation to scientific knowledge, employ men who pursue science, provided they are also men of business. This is perhaps now done to some extent, but to nothing like the degree which is practicable with advantage to all parties. The right means cannot have occurred to a Government which has not yet learned to approach and distinguish the class as a whole.

"At the same time, I am free to confess that I am unable to advise how that which I think should be may come to pass. I believe I have written the expression of feelings rather than the conclusions of judgment, and I would wish your lordship to consider this letter as private rather than as one addressed to the chairman of a committee.

　　　　"I have the honour to be, my Lord,
　　　　　　"Your very faithful Servant,
　　　　　　　　"M. FARADAY."

In this day, when so many allow their names to be used for offices of which they never intended to discharge the duties, the following letter may convey an appropriate lesson :—

　　　　　　　　　　　　"ROYAL INSTITUTION, *Oct.* 17*th*, 1849.
"MY DEAR PERCY,

　　　"I cannot be on the committee ; I avoid everything of that kind, that I may keep my stupid mind a little clear. As to being on a committee and not working, that is worse still. * * *

　　　　　　　"Ever yours and Mrs. Percy's,

　　　　　　　　　　　　"M. FARADAY."

It is well known that he waged implacable war with the Spiritualists. Eighteen years ago tables took to spinning mysteriously under the fingers of ladies and gentlemen who sat or stood around the animated furniture ; much was said about a new force, much too about strange revelations from another sphere, but Faraday made a simple apparatus which convinced him and most others that the tables moved through the unconscious pressure of the hands that touched them. The account of this will be found in the *Athenæum* of July 2, 1853. Three weeks afterwards he wrote to his friend Schönbein : " I have not been at work except in turning the tables upon the table-turners, nor should I have done that, but that so many inquiries poured in upon me, that I thought it better to stop the inpouring flood by letting all know at once what my views and thoughts were. What a weak, credulous, incredulous, unbelieving, superstitious, bold, frightened,—what a ridiculous world ours is, as far as concerns the mind of man. How full of inconsistencies, contradictions, and absurdities it is ! " But the believers in

　　　　　　　　　　　I

these occult phenomena, some of them holding high positions about the Court, would not let him alone; and there
are many indications of the annoyance and irritation they
caused him. He declined to meet the professors of the
mysterious art, and the following letter will serve to show
the way in which he regarded them :—

"ROYAL INSTITUTION, *Nov.* 1, 1864.
"SIR,

"I beg to thank you for your papers, but have wasted
more thought and time on so-called spiritual manifestation
than it has deserved. Unless the spirits are utterly con-
temptible, *they* will find means to draw my attention.

"How is it that your name is not signed to the testimony
that you give? Are you doubtful even whilst you publish?
I've no evidence that any natural or unnatural power is
concerned in the phenomena that requires investigation or
deserves it. If I could consult the spirits, or move them to
make themselves honestly manifest, I would do it. But I
cannot, and am weary of them.

"I am, Sir, your obedient Servant,
"M. FARADAY."

There was once a strange statement put forth to the effect
that Faraday said electricity was life.[1] He himself denied it
indignantly; but as most falsehoods are perversions of some
truth, this one probably originated in his experiments on the
Gymnotus. He felt an intense interest in those marine
animals that give shocks, and sought "to identify the living

---

[1] I myself once heard this advanced by an infidel lecturer on Pad-
dington Green.

power which they possess, with that which man can call into action from inert matter, and by him named electricity." [1] The most powerful of these is the Gymnotus, or electrical eel, and a live specimen of this creature, forty inches long, was secured by the Adelaide Gallery—a predecessor of the Polytechnic—in the summer of 1838. Four days after its arrival the poor creature lost an eye; for two months it could not be coaxed to eat either meat or fish, worms or frogs; but at last one day it killed and devoured four small fishes, and afterwards swallowed about a fish per diem. It was accustomed to swim round and round the tank, till a live fish was dropped in, when, in some cases bending round its victim, it would discharge a shock that made the fish float on its back stunned and ready to be sucked into the jaws of its assailant.

Faraday examined this eel and the water around it, both with his hands and with special collectors of electricity, and satisfied himself not merely of the shock, which was easy enough, but of its power to deflect a galvanometer, to make a magnet, to effect chemical decomposition, and to give a spark. His account of the experiments terminates with some speculations on the connection of this animal electricity with nervous power; but there the matter rested. His own views were thus expressed to his friend Dumas :—" As living creatures produce heat, and a heat certainly identical with that of our hearths, why should they not produce electricity also, and an electricity in like manner identical with that of our machines? But if the heat produced during life, and necessary to life, is not life after all, why should electricity

[1] " Electrical Researches," Series XV.

itself be life? Like heat, like chemical action, electricity is
an implement of life, and nothing more."

Whether the belief that electricity is life would be incon-
sistent with the Christian faith or not, it is clear that when
an infidel preacher asserts that Faraday held such an opinion,
his assertion will influence few who are not already disposed
to materialism. Far more damaging is it to the cause of
religion when her ministers repeat the assumption of the
infidel that those who study the truths of nature are particularly prone to disbelieve. Yet such statements have
been made, even with reference to Faraday. I have it on
the best authority that one of the leading clergymen of the
day, preaching on a special occasion from Peter's words,
"The elements shall melt with fervent heat, the earth also
and the works that are therein shall be burned up," spoke
in antagonism to scientific men, alluding to Faraday by
name, and to his computation of the tremendous electrical
forces that would be produced by sundering the elements of
one drop of water. "They shall be confuted by their own
element—fire," added the preacher, careless of the conclusion which his audience might legitimately draw from
such a two-edged argument. The accuser of the men of
science was much astonished when told after his sermon,
by a brother clergyman, that Faraday and other eminent
physicists of the day were believers in a Divine revelation.

It may be doubted whether Faraday ever tried to form a
definite idea of the relation in which the physical forces
stand to the Supreme Intelligence, as Newton did, or his
own friend Sir John Herschel; nor did he consider it part
of his duty as a lecturer to look beyond the natural laws he

was describing. His practice in this respect has been well described by the Rev. Professor Pritchard : [1]—"This great and good man never obtruded the strength of his faith upon those whom he publicly addressed ; upon principle he was habitually reticent on such topics, because he believed they were ill suited for the ordinary assemblages of men. Yet on more than one occasion when he had been discoursing on some of the magnificent pre-arrangements of Divine Providence so lavishly scattered in nature, I have seen him struggle to repress the emotion which was visibly striving for utterance; and then, at the last, with one single far-reaching word, he would just hint at his meaning rather than express it. On such occasions he only who had ears to hear, could hear."

In his more familiar lectures to the cadets at Woolwich, however, he more than hinted at such elevated thoughts. In conversation, too, Faraday has been known to express his wonder that anyone should fail to recognize the constant traces of design ; and in his writings there sometimes occur such passages as the following :—"When I consider the multitude of associated forces which are diffused through nature—when I think of that calm and tranquil balancing of their energies which enables elements most powerful in themselves, most destructive to the world's creatures and economy, to dwell associated together and be made subservient to the wants of creation, I rise from the contemplation more than ever impressed with the wisdom, the beneficence, and grandeur beyond our language to express, of the Great Disposer of all !"

Faraday's journals abound with descriptions of "nature

[1] "Analogies in the Progress of Nature and Grace," p. 121.

and human nature." He had evidently a keen eye for the beauties of scenery, and occasionally the objects around him suggested higher thoughts. Here are two instances taken from his notes of a Swiss tour in 1841 :—

"*Monday*, 19*th*.—Very fine day; walk with dear Sarah on the lake side to Oberhofen, through the beautiful vineyards; very busy were the women and men in trimming the vines, stripping off leaves and tendrils from the fruit-bearing branches. The churchyard was beautiful, and the simplicity of the little remembrance-posts set upon the graves very pleasant. One who had been too poor to put up an engraved brass plate, or even a painted board, had written with ink on paper the birth and death of the being whose remains were below, and this had been fastened to a board, and mounted on the top of a stick at the head of the grave, the paper being protected by a little edge and roof. Such was the simple remembrance, but Nature had added her pathos, for under the shelter by the writing a caterpillar had fastened itself, and passed into its deathlike state of chrysalis, and, having ultimately assumed its final state, it had winged its way from the spot, and had left the corpse-like relics behind. How old and how beautiful is this figure of the resurrection ! Surely it can never appear before our eyes without touching the thoughts."

"*August* 12*th*, *Brienz Lake.*—George and I crossed the lake in a boat to the Giessbach—he to draw, and I to saunter. . . . This most beautiful fall consists of a fine river, which passes by successive steps down a very deep precipice into the lake. In some of these steps there is a clear leap of water of 100 feet or more, in others most beautiful combinations of leap, cataract, and rapid, the finest rocks occurring at the sides and bed of the torrent. In one part a bridge passes over it. In another a cavern and a path occur under it. To-day every fall was foaming from the abundance of water, and the current of wind brought down by it was in some parts almost too strong to stand against. The sun shone brightly, and the rainbows seen from various points were very beautiful. One at the bottom of a fine but furious fall was very pleasant. There it remained motion-less, whilst the gusts and clouds of spray swept furiously across its place, and were dashed against the rock. It looked like a spirit strong in faith and steadfast in the midst of the storm of passions sweeping across it; and though it might fade and revive, still it held on to the rock as in hope and giving hope; and the very drops which in the whirlwind of their fury seemed as if they would carry all away, were made to revive it and give it greater beauty.

"How often are the things we fear and esteem as troubles made to become blessings to those who are led to receive them with humility and patience."

In concluding this section it may be well to string to-gether a few gems from Faraday's lectures or correspondence, though they are greatly damaged by being torn away from their original setting :—

"After all, though your science is much to me, we are not friends for science sake only, but for something better in a man, something more important in his nature, affec-tion, kindness, good feeling, moral worth; and so, in remembrance of these, I now write to place myself in your presence, and in thought shake hands, tongues, and hearts together." This was addressed to Schönbein.

"I should be glad to think that high mental powers insured something like a high moral sense, but have often been grieved to see the contrary; as also, on the other hand, my spirit has been cheered by observing in some lowly and uninstructed creature such a healthful and honourable and dignified mind as made one in love with human nature. When that which is good mentally and morally meet in one being, that that being is more fitted to work out and manifest the glory of God in the creation, I fully admit."

"Let me, as an old man who ought by this time to have profited by experience, say that when I was younger I found I often misinterpreted the intentions of people, and found they did not mean what at the time I supposed they meant; and further, that as a general rule, it was better to be a little dull of apprehension when phrases seemed to imply pique, and quick in perception when, on the contrary, they seemed to imply kindly feeling. The real truth never fails ultimately to appear; and opposing parties, if wrong, are sooner convinced when replied to forbearingly, than when overwhelmed."

"Man is an improving animal. Unlike the animated world around him, which remains in the same constant state, he is continually varying; and it is one of the noblest prerogatives of his nature, that in the highest of earthly distinctions he has the power of raising and exalting himself continually. The transitory state of man has been

held up to him as a memento of his weakness : to man
degraded it may be so with justice ; to man as he ought
to be it is no reproach ; and in knowledge, that man only
is to be contemned and despised who is *not* in a state
of transition."

"It is not the duty or place of a philosopher to dictate belief, and all hypothesis is more or less matter of
belief; he has but to give his facts and his conclusions,
and so much of the logic which connects the former with
the latter as he may think necessary, and then to commit
the whole to the scientific world for present, and, as he
may sometimes without presumption believe, for future
judgment."

SECTION IV.

HIS METHOD OF WORKING.

IT is on record that when a young aspirant asked Faraday the secret of his success as a scientific investigator, he replied, " The secret is comprised in three words—Work, Finish, Publish."

Each of these words, we may be sure, is full of meaning, and will guide us in a useful inquiry.

Already in the " Story of his Life " we have caught some glimpses of the philosopher at work in his laboratory; but before looking at him more closely let us learn from a foreigner with what feelings to enter a place that is hallowed by so many memories sacred in the history of science. Professor Schönbein, of Basle, who visited England in 1840, says : " During my stay in London, I once worked with Faraday for a whole day long in the laboratory of the Royal Institution, and I cannot forbear to say that this was one of the most enjoyable days that I ever spent in the British Capital. We commenced our day's work with break-fast; and when that was over I was supplied with one of the laboratory dresses of my friend, which, when I was pre-

sented in it to the ladies, gave occasion to no little amusement, as the dimensions of Faraday are different from those
of my precious body.

" To work with a man like Faraday was in itself a great
pleasure; but this pleasure was not a little heightened in
doing so in a place where such grand secrets of nature had
been unfolded, the most brilliant discoveries of the century
had been made, and entirely new branches of knowledge
had been brought forth. For the empty intellect circumstances of this nature are indeed of little special value; but
they stand in quite another relation to our power of imagination and inner nature.

" I do not deny that my surroundings produced in me
a very peculiar feeling; and whilst I trod the floor upon
which Davy had once walked—whilst I availed myself of
some instrument which this great discoverer had himself
handled—whilst I stood working at the very table at which
the ever-memorable man sought to solve the most difficult
problems of science, at which Faraday enticed the first
sparks out of the magnet, and discovered the most beautiful laws of the chemical action of current electricity,
I felt myself inwardly elevated, and believed that I myself
experienced something of the inbreathing of the scientific
spirit which formerly ruled there with such creative power,
and which still works on." [1]

The habit of Faraday was to think out carefully before-
hand the subject on which he was working, and to plan
his mode of attack. Then, if he saw that some new piece

[1] " Mittheilungen aus dem Reisetagebuche eines deutschen Natur-
forschers," p. 275.

of apparatus was needed, he would describe it fully to the instrument maker with a drawing, and it rarely happened that there was any need of alteration in executing the order. If, however, the means of experiment existed already, he would give Anderson a written list of the things he would require, at least a day before—for Anderson was not to be hurried. When all was ready, he would descend into the laboratory, give a quick glance round to see that all was right, take his apron from the drawer, and rub his hands together as he looked at the preparations made for his work. There must be no tool on the table but such as he required. As he began, his face would be exceedingly grave, and during the progress of an experiment all must be perfectly quiet; but if it was proceeding according to his wish, he would commence to hum a tune, and sometimes to rock himself sideways, balancing alternately on either foot. Then, too, he would often talk to his assistant about the result he was expecting. He would put away each tool in its own place as soon as done with, or at any rate when the day's work was over, and he would not unnecessarily take a thing away from its place : thus, if he wanted a per-forated cork, he would go to the drawer which contained the corks and cork-borers, make there what he wanted, replace the borers, and shut the drawer. No bottle was allowed to remain without its stopper ; no open glass might stand for a night without a paper cover ; no rubbish was to be left on the floor ; bad smells were to be avoided if pos-sible ; and machinery in motion was not permitted to grate. In working, also, he was very careful not to employ more force than was wanted to produce the effect. When his experiments were finished and put away, he would leave the laboratory, and think further about them upstairs.

This orderliness and this economy of means he not only practised himself, but he expected them also to be followed by any who worked with him; and it is from conversation with these that I have been enabled to give this sketch of his manner of working.

This exactness was also apparent in the accounts he kept with the Royal Institution and Trinity House, in which he entered every little item of expenditure with the greatest minuteness of detail.

It was through this lifelong series of experiments that Faraday won his knowledge and mastered the forces of nature. The rare ingenuity of his mind was ably seconded by his manipulative skill, while the quickness of his perceptions was equalled by the calm rapidity of his movements.

He had indeed a passion for experimenting. I recollect his meeting me at the entrance to the lecture theatre at Jermyn Street, when Lyon Playfair was to give the first, or one of the first lectures ever delivered in the building. "Let us go up here," said he, leading me far away from the central table. I asked him why he chose such an out-of-the-way place. "Oh," he replied, "we shall be able here to find out what are the acoustic qualities of the room."

The simplicity of the means with which he made his experiments was often astonishing, and was indeed one of the manifestations of his genius.

A good instance is thus narrated by Sir Frederick Arrow. " When the electric light was first exhibited permanently at

Dungeness, on 6th June, 1862, a committee of the Elder
Brethren, of which I was one, accompanied Faraday to
observe it: We dined, I think, at Dover, and embarked in
the yacht from there, and were out for some hours watching
it, to Faraday's great delight—(a very fine night),—and
especially we did so from the Varne lightship, about equi-
distant between it and the French light of Grisnez, using
all our best glasses and photometers to ascertain the relative value of the lights : and this brings me to my story.
Before we left Dover, Faraday, with his usual bright smile,
in great glee showed me a little common paper box, and
said, ' I must take care of this ; it's my special photometer '—and then, opening it, produced a lady's ordinary
black shawl-pin, — jet, or imitation perhaps, — and then
holding it a little way off the candle, showed me the image
very distinct ; and then, putting it a little further off, placed
another candle near it, and the relative distance was shown
by the size of the image. He lent me this afterwards when
we were at the Varne lightship, and it acted admirably ;
and ever since I have used one as a very convenient mode
of observing, and I never do so but I think of that night
and dear good Faraday, and his genial happy way of
showing how even common things may be made useful."
After this Faraday modified his glass-bead photometer, and
he might be seen comparing the relative intensity of two
lights by watching their luminous images on a bead of
black glass, which he had threaded on a string, and was
twirling round so as to resolve the brilliant points into
circles of fainter light ; or he fixed the black glass balls on
pieces of cork, and, attaching them to a little wheel, set

them spinning for the same purpose. Some of these beads are preserved by the Trinity House, with other treasures of a like kind, including a flat piece of solder of an irregular oval form, turned up at one side so as to form a thumb-rest, and which served the philosopher as a candlestick to support the wax-light that he used as a standard. The museum of the Royal Institution contains a most instructive collection of his experimental apparatus, including the common electrical machine which he made while still an apprentice at Riebau's, and the ring of soft iron, with its twisted coils of wire isolated by calico and tied with common string, by means of which he first obtained electrical effects from a magnet.

A lady, calling on his wife, happened to mention that a needle had been once broken into her foot, and she did not know whether it had been all extracted or not. " Oh !" said Faraday, " I will soon tell you that,"—and taking a finely suspended magnetic needle, he held it close to her foot, and it dipped to the concealed iron.

"An artist was once maintaining that in natural appearances and in pictures, up and down, and high and low, were fixed indubitable realities; but Faraday told him that they were merely conventional acceptations, based on standards often arbitrary. The disputant could not be convinced that ideas which he had hitherto never doubted had such shifting foundations. 'Well,' said Faraday, 'hold a walking-stick between your chin and your great toe; look along it and say which is the upper end.' The experiment was tried, and the artist found his idea of perspective at complete variance with his sense of reality; either end of the stick might be called 'upper,'—pictorially it was one, physically it was the other."

On this subject Schönbein has also some good remarks. "The laboratory of the Institution is indeed efficiently arranged, though anything but large and elaborately furnished. And yet something extraordinary has happened in this room for the extension of the limits of knowledge; and already more has been done in it than in many other institutions where the greatest luxury in the supply of apparatus prevails, and where there is the greatest command of money. But when men work with the creative genius of a Davy, and the intuitive spirit of investigation and the wealth of ideas of a Faraday, important and great things must come to pass, even though the appliances at command should be of so limited a character. For the experimental investigator of nature, it is especially desirable that, according to the kind of his researches, he should have at command such and such appliances, that he should possess a 'philosophical apparatus,' a laboratory, &c.; but for the purpose of producing something important, of greatly widening the sphere of knowledge, it in no way follows that a superfluity of such things is necessary to him. . . . He who understands how to put appropriate questions to Nature, generally knows how to extract the answers by simple means; and he who wants this capacity will, I fear, obtain no profitable result, even though all conceivable tools and apparatus may be ready to his hand."

Nor did Faraday require elaborate apparatus to illustrate his meaning. Steaming up the Thames one July day in a penny boat, he was struck with the offensiveness of the

water.   He tore some white cards into pieces, wetted them
so as to make them sink easily, and dropped them into the
river at each pier they came to.   Their sudden disappearance
from sight, though the sun was shining brightly, was proof
enough of the impurity of the stream ; and he wrote a letter
to the *Times* describing his observations, and calling public
attention to the dangerous state of the river.[1]   At a meeting
of the British Association he wished to explain the manner
in which certain crystallized bodies place themselves between
the poles of an electro-magnet : two or three raw potatoes
furnished the material out of which he cut admirable models
of the crystals.

Faraday's manner of experimenting may be further illustrated by the recollections of other friends who have had
the opportunity of watching him at work.

Mr. James Young, who was in the laboratory of University College in 1838, thus writes :—"About that time
Professor Graham had got from Paris Thilorier's apparatus
for producing liquid and solid carbonic acid ; hearing of
this, Mr. Faraday came to Graham's laboratory, and, as one
might expect, showed great interest in this apparatus, and
asked Graham for the loan of it for a Friday evening lecture at the Royal Institution, which of course Graham
readily granted, and Faraday asked me to come down to
the Institution and give him the benefit of my experience
in charging and working the apparatus ; so I spent a long
evening at the Royal Institution laboratory.   There was

---

[1] *Punch's* cartoon next week represented Professor Faraday holding
his nose, and presenting his card to Father Thames, who rises out of the
unsavoury ooze.

K

no one present but Faraday, Anderson, and myself. The principal thing we did was to charge the apparatus and work with the solid carbonic acid, Mr. Faraday working with great activity: his motions were wonderfully rapid; and if he had to cross the laboratory for anything, he did not walk at an ordinary step, but ran for it, and when he wanted anything he spoke quickly. Faraday had a theory at that time that all metals would become magnetic if their temperature were low enough; and he tried that evening some experiments with cobalt and manganese, which he cooled in a mixture of carbonic acid and ether, but the results were negative."

Among the deep mines of the Durham coalfield is one called the Haswell Colliery. One Saturday afternoon, while the men were at work in it as usual, a terrible explosion occurred: it proceeded from the fire-damp that collects in the vaulted space that is formed in old workings when the supporting pillars of coal are removed and the roof falls in: the suffocating gases rushed along the narrow passages, and overwhelmed ninety-five poor fellows with destruction. Of course there was an inquiry, and the Government sent down to the spot as their commissioners Professor Faraday and Sir Charles Lyell. The two gentlemen attended at the coroner's inquest, where they took part in the examination of the witnesses; they inspected the shattered safety-lamps; they descended into the mine, spending the best part of a day in the damaged and therefore dangerous galleries where the catastrophe had occurred, and they did not leave without showing in a practical form their sympathy with the sufferers. When down in the

pit, an inspector showed them the way in which the work-
men estimated the rapidity of the ventilation draught, by
throwing a pinch of gunpowder through the flame of a
candle, and timing the movement of the little puff of
smoke. Faraday, not admiring the free and easy way in
which they handled their powder, asked where they kept
their store of it, and learnt that it was in a large black
bag which had been assigned to him as the most comfortable seat they could offer. We may imagine the liveliness
with which he sprang to his feet, and expostulated with
them on their culpable carelessness.

My own opportunities of observing Faraday at work were
nearly confined to a series of experiments, which are the
better worth describing here as they have escaped the
notice of previous biographers. The Royal Commission
appointed to inquire into our whole system of Lights,
Buoys, and Beacons, perceived a great defect that rendered many of our finest shore or harbour lights comparatively ineffective. The great central lamp in a lighthouse is surrounded by a complicated arrangement of
lenses and prisms, with the object of gathering up as many
of the rays as possible and sending them over the surface
of the sea towards the horizon. Now, it is evident that
if this apparatus be adjusted so as to send the beam two
or three degrees upwards, the light will be lost to the ship-
ping and wasted on the clouds, and if two or three degrees
downwards, it will only illuminate the water in the neighbourhood : in either case the beautiful and expensive
apparatus would be worse than useless. It is evident also
that if the eye be placed just above the wick of the lamp,

it will see through any particular piece of glass that very portion of the landscape which will be illuminated by a ray starting from the same spot; or the photographic image formed in the place of the flame by any one of the lenses will tell us the direction in which that lens will throw the luminous rays. This simple principle was applied by the Commissioners for testing the adjustment of the apparatus in the different lights, and it was found that few were rightly placed, or rather that no method of adjustment was in use better than the mason's plumbline. The Royal Commissioners therefore in 1860 drew the attention of all the lighthouse authorities to this fact, and asked the Elder Brethren of the Trinity House, with Faraday and other parties, to meet them at the lights recently erected at the North Foreland' and Whitby. I, as the scientific member of the Commission, had drawn out in detail the course of rays from different parts of the flame, through different parts of the apparatus, and I was struck with the readiness with which Faraday, who had never before considered the matter,[1] took

[1] Since writing the above I have come across a letter written by Faraday in answer to one by Captain Weller as far back as 13th Sept. 1839, in which he pointed out the maladjustment of the dioptric apparatus at Orfordness. In July of the following year he made lengthy suggestions to the Trinity House, in which he proposed using a flat white circle or square, half an inch across, on a piece of black paper or card, as a "focal object." This was to be looked at from outside, in order to test the regularity of the glass apparatus. He also suggested observations on the divergence by looking at this white circle at a distance of twenty feet at most. Another plan he proposed was that of lighting the lamp and putting up a white screen outside. These methods of examining he carried out very shortly afterwards at Blackwall, on French and English refractors, but it seems never to have occurred to

up the idea, and recognized its importance and its practical application. With his characteristic ingenuity, too, he devised a little piece of apparatus for the more exact observation of the matter inside the lighthouse. He took to Mr. Ladd, the optical instrument maker, a drawing, very neatly executed, with written directions, and a cork cut into proper shape with two lucifer matches stuck through it, to serve as a further explanation of his meaning : and from this the " focimeter," as he called it, was made. The position of the glass panels at Whitby was corrected by means of this little instrument, and there were many journeys down to Chance's glassworks near Birmingham, where, declining the hospitality of the proprietor in order to be absolutely independent, he put up at a small hotel while he made his experiments, and jotted down his observations on the cards he habitually carried in his pocket. At length we were invited down to see the result. Faraday explained carefully all that had been done, and at the risk of sea-sickness (no trifling matter in his case) accompanied us out to sea to observe the effect from various directions and at various distances. The experience acquired at Whitby was applied elsewhere, and in May 1861 the Trinity House appointed a Visiting Committee, " to examine all dioptric light establishments, with the view of remedying any inaccuracies of arrangement that may be found to exist." Faraday had instructed and practised Captain Nisbet and some others of the Elder Brethren in the use of the foci-meter, and now wrote a careful letter of suggestions on the question of adjustment between the lamp and the lenses

him to place his eye in the focus, or in any other manner to observe the course of the rays from inside the apparatus.

and prisms; so thoughtfully did he work for the benefit of those who "go down to the sea in ships, that do business in great waters."

As to the mental process that devised, directed, and interpreted his experiments, it must be borne in mind that Faraday was no mathematician; his power of appreciating an *à priori* reason often appeared comparatively weak. "It has been stated on good authority that Faraday boasted on a certain occasion of having only once in the course of his life performed a mathematical calculation: that once was when he turned the handle of Babbage's calculating machine."[1] Though there was more pleasantry than truth in this professed innocence of numbers, probably no one acquainted with his electrical researches will doubt that, had he possessed more mathematical ability, he would have been saved much trouble, and would sometimes have expressed his conclusions with greater ease and precision. Yet, as Sir William Thomson has remarked with reference to certain magnetic phenomena, "Faraday, without mathematics, divined the result of the mathematical investigation; and, what has proved of infinite value to the mathematicians themselves, he has given them an articulate language in which to express their results. Indeed, the whole language of the magnetic field and 'lines of force' is Faraday's. It must be said for the mathematicians that they greedily accepted it, and have ever since been most zealous in using it to the best advantage."

The peculiarity of his mind was indeed well known to himself. In a letter to Dr. Becker he says: "I was

[1] Dr. Scoffern, *Belgravia*, October 1867.

never able to make a fact my own without seeing it; and
the descriptions of the best works altogether failed to
convey to my mind such a knowledge of things as to
allow myself to form a judgment upon them. It was so
with *new* things. If Grove, or Wheatstone, or Gassiot, or
any other told me a new fact, and wanted my opinion
either of its value, or the cause, or the evidence it could
give on any subject, I never could say anything until I
had seen the fact. For the same reason I never could
work, as some Professors do most extensively, by students
or pupils. All the work had to be my own."

   The following story by Mr. Robert Mallet serves as an
illustration :—" It must be now eighteen years ago when I
paid him a visit and brought some slips of flexible and
tough Muntz's yellow metal, to show him the instantaneous
change to complete brittleness with rigidity produced by
dipping into pernitrate of mercury solution. He got the
solution, and I *showed* him the facts ; he obviously did not
doubt what he saw *me* do before and close to him : but a
sort of experimental instinct seemed to require he should try
it himself. So he took one of the slips, bent it forwards
and backwards, dipped it, and broke it up into short bits between his own fingers. He had not before spoken. *Then*
he said, ' Yes, it *is* pliable, and it *does* become instantly
brittle.' And after a few moments' pause he added, ' Well,
now have you any more facts of the sort ? ' and seemed
a little disappointed when I said ' No ; none that are new.'
It has often since occurred to me how his mind needed
absolute satisfaction that he had grasped a *fact*, and then
instantly rushed to colligate it with another if possible."

But as the Professor watched these new facts, new
thoughts would shape themselves in his mind, and this
would lead to fresh experiments in order to test their truth.
The answers so obtained would lead to further questions.
Thus his work often consisted in the defeat of one hypothesis after another, till the true conditions of the phenomena
came forth and claimed the assent of the experimenter and
ultimately of the scientific world.

A. de la Rive has some acute observations on this
subject. He explains how Faraday did not place himself
before his apparatus, setting it to work, without a preconceived idea.   Neither did he take up known phenomena,
as some scientific men do, and determine their numerical
data, or study with great precision the laws which regulate
them. " A third method, very different from the preceding,
is that which, quitting the beaten track, leads, as if
by inspiration, to those great discoveries which open new
horizons to science. This method, in order to be fertile,
requires one condition—a condition, it is true, which is
but rarely met with—namely, genius. Now, this condition
existed in Faraday.   Endowed, as he himself perceived,
with much imagination, he dared to advance where many
others would have recoiled : his sagacity, joined to an
exquisite scientific tact, by furnishing him with a presenti-
ment of the possible, prevented him from wandering into
the fantastic ; while, always wishing only for facts, and
accepting theories only with difficulty, he was nevertheless more or less directed by preconceived ideas, which,
whether true or false, led him into new roads, where most
frequently he found what he sought, and sometimes also

what he did not seek, but where he constantly met with some important discovery.

" Such a method, if indeed it can be called one, although barren and even dangerous with mediocre minds, produced great things in Faraday's hands ; thanks, as we have said, to his genius, but thanks also to that love of truth which characterized him, and which preserved him from the temptation so often experienced by every discoverer, of seeing what he wishes to see, and not seeing what he dreads."

This love of truth deserves a moment's pause. It was one of the most beautiful and most essential of his characteristics; it taught him to be extremely cautious in receiving the statements of others or in drawing his own conclusions,[1]

---

[1] A good instance of his caution in drawing conclusions is contained in one of his letters to me :—

" ROYAL INSTITUTION OF GREAT BRITAIN,
2 *July*, 1859.

" MY DEAR GLADSTONE,

" Although I have frequently observed lights from the sea, the only thing I have learnt in relation to their *relative brilliancy* is that the average of a very great number of observations would be required for the attainment of a moderate approximation to truth.   One has to be some miles off at sea, or else the observation is not made in the chief ray, and then one does not know the state of the atmosphere about a given lighthouse.   Strong lights like that of Cape Grisnez have been invisible when they should have been strong ; feeble lights by comparison have risen up in force when one might have expected them to be relatively weak ; and after inquiry has not shown a state of the air at the lighthouse explaining such differences.   It is probable that the cause of difference often exists at sea.

" Besides these difficulties there is that other great one of not seeing the two lights to be compared in the field of view at the same time and same distance.   If the eye has to turn 90° from one to the other, I have no confidence in the comparison ; and if both be in the field

and it led him, if his scepticism was overcome, to adopt at once the new view, and to maintain it, if need be, against the world.

" The thing I am proudest of, Pearsall, is that I have never been found to be wrong," he could say in the early part of his scientific history without fear of contradiction. After his death A. de la Rive wrote, " I do not think that Faraday has once been caught in a mistake ; so precise and conscientious was his mode of experimenting and observing." This is not absolutely true ; but the extreme rarity of his mistakes, notwithstanding the immense amount of his published researches, is one of those marvels which can be appreciated only by those who are in the habit of

of sight at once, still unexpected and unexplained causes of difference occur. The two lights at the South Foreland are beautifully situated for comparison, and yet sometimes the upper did not equal the lower when it ought to have surpassed it. This I referred at the time to an upper stratum of haze ; but on shore they knew nothing of the kind, nor had any such or other reason to expect particular effects.

" Ever truly yours,

" M. FARADAY."

As an instance of his unwillingness to commit himself to an opinion unless he was sure about it, may be cited a letter he wrote to Mr. Airy, the Astronomer Royal, who asked for his advice in regard to the material of which the national standard of length should be made :—
" I do not see any reason why a pure metal should be particularly free from internal change of its particles, and on the whole should rather incline to the hard alloy than to soft copper, and yet I hardly know why. I suppose the labour would be too great to lay down the standard on different metals and substances ; and yet the comparison of them might be very important hereafter, for twenty years seem to *do* or *tell* a great deal in relation to standard measures." Bronze was finally chosen.

describing what they have seen in the mist land that lies beyond the boundaries of previous knowledge.

Into this unknown region his mental vision was ever stretched. " I well remember one day," writes Mr. Barrett, a former assistant at the Royal Institution, " when Mr. Faraday was by my side, I happened to be steadying, by means of a magnet, the motion of a magnetic needle under a glass shade. Mr. Faraday suddenly looked most impressively and earnestly as he said, ' How wonderful and mysterious is that power you have there ! the more I think over it the less I seem to know : '—and yet he who said this knew more of it than any living man."

It is easy to imagine with what wonder he would stand before the apples or leaves or pieces of meat that swung round into a transverse position between the poles of his gigantic magnet, or the sand that danced and eddied into regular figures on plates of glass touched by the fiddle-bow, or gold so finely divided that it appeared purple and when diffused in water took a twelvemonth to settle. It is easy, too, to imagine how he would long to gain a clear idea of what was taking place behind the phenomena. But it is far from easy to grasp the conceptions of his brain : language is a clumsy vehicle for such thoughts. He strove to get rid of such figurative terms as " currents " and " poles "; in discussing the mode of propagation of light and radiant heat he endeavoured " to dismiss the ether, but not the vibrations "; and in conceiving of atoms, he says : " As to the little solid particles . . . I cannot form any idea of them apart from the forces, so I neither admit nor deny them. They do not afford me the least help in

my endeavour to form an idea of a particle of matter. On
the contrary, they greatly embarrass me." Yet he could not
himself escape from the tyranny of words or the deceitful-
ness of metaphors, and it is hard for his readers to comprehend what was his precise idea of those centres of forces
that occupy no space, or of those lines of force which he
beheld with his mental eye, curving alike round his magnetic needle, and that mightiest of all magnets—the earth.

As he was jealous of his own fame, and had learnt by
experience that discoveries could be stolen, he talked little
about them till they were ready for the public ; indeed, he
has been known to twit a brother electrician for telling his
discoveries before printing them, adding with a knowing
laugh, "I never do that." He was obliged, however, to
explain his results to Professor Whewell, or some other
learned friend, if he wished to christen some new idea with
a Greek name. One of Whewell's letters on such an occasion, dated Trinity College, Cambridge, October 14, 1837,
begins thus :—

" MY DEAR SIR,
         " I am always glad to hear of the progress of your
researches, and never the less so because they require the
fabrication of a new word or two. Such a coinage has
always taken place at the great epochs of discovery ; like
the medals that are struck at the beginning of a new
reign, or rather like the change of currency produced by
the accession of a new Sovereign ; for their value and influence consists in their coming into common circulation."

*   *   *   *   *   *

During the whole time of an investigation Faraday had kept ample notes, and when all was completed he had little to do but to copy these notes, condensing or re-arranging some parts, and omitting what was useless. The paper then usually consisted of a series of numbered paragraphs, containing first a statement of the subject of inquiry, then a series of experiments giving negative results, and afterwards the positive discoveries. In this form it was sent to the Royal Society or some other learned body. Yet this often involved considerable labour, as the following words written to Miss Moore in 1850 from a summer retreat in Upper Norwood will show :—" I write and write and write, until nearly three papers for the Royal Society are nearly completed, and I hope that two of them will be good if they do justify my hopes, for I have to criticise them again and again before I let them loose. You shall hear of them at some of the next Friday evenings."

This criticism did not cease with their publication, for he endeavoured always to improve on his previous work. Thus, in 1832 he bound his papers together in one volume, and the introduction on the fly-leaf shows the object with which it was done :—

" Papers of mine, published in octavo, in the *Quarterly Journal of Science*, and elsewhere, since the time that Sir H. Davy encouraged me to write the analysis of caustic lime.

" Some, I think (at this date), are good, others moderate, and some bad. But I have put *all* into the volume, because of the utility they have been of to me—and none more than the bad—in pointing out to me in future, or rather after times, the faults it became me to watch and to avoid.

"As I never looked over one of my papers a year after it was written, without believing, both in philosophy and manner, it could have been much better done, I still hope the collection may be of great use to me.

"·M. FARADAY.

"*August* 18, 1832."

This section may be summed up in the words of Dumas when he gave the first "Faraday Lecture" of the Chemical Society :—"Faraday is the type of the most fortunate and the most accomplished of the learned men of our age. His hand in the execution of his conceptions kept pace with his mind in designing them; he never wanted boldness when he undertook an experiment, never lacked resources to ensure success, and was full of discretion in interpreting results. His hardihood, which never halted when once he had undertaken a task, and his wariness, which felt its way carefully in adopting a received conclusion, will ever serve as models for the experimentalist."

SECTION V.

THE VALUE OF HIS DISCOVERIES.

SCIENCE is pursued by different men from different motives.

> " To some she is the goddess great ;
>   To some the milch-cow of the field :
> Their business is to calculate
>   The butter she will yield."

Now, Faraday had been warned by Davy before he entered his service that Science was a mistress who paid badly; and in 1833 we have seen him deliberately make his calculation, give up the butter, and worship the goddess.

For the same reason also he declined most of the positions of honour which he was invited to fill, believing that they would encroach too much on his time, though he willingly accepted the honorary degrees and scientific distinctions that were showered upon him.[1]

---

[1] De la Rive points this out in his brief notice of Faraday immediately on receiving the news of his death :—" Je n'ai parlé que du savant, je tiens aussi à dire un mot de l'homme. Alliant à une modestie vraie, parcequ'elle provenait de l'élévation de son âme, une droiture à toute épreuve et une candeur admirable, Faraday n'aimait la science que pour elle-même. Aussi jouissait-il des succès des autres au moins autant

And among those who follow Science lovingly, there are two very distinct bands : there are the philosophers, the discoverers, men who persistently ask questions of Nature ; and there are the practical men, who apply her answers to the various purposes of human life.   Many noble names are inscribed in either bead-roll, but few are able to take rank in both services : indeed, the question of practical utility would terribly cramp the investigator, while the enjoyment of patient research in unexplored regions of knowledge is usually too ethereal for those who seek their pleasure in useful inventions.   The mental configuration is different in the two cases ; each may claim and receive his due award of honour.

Faraday was pre-eminently a discoverer; he liked the name of "philosopher."   His favourite paths of study seem to wander far enough from the common abodes of human thought or the requirements of ordinary life.   He became familiar, as no other man ever was, with the varied forces of magnetism and electricity, heat and light, gravitation and galvanism, chemical affinity and mechanical motion ; but he did not seek to "harness the lightnings," or to chain those giants and make them grind like Samson in the prison-house.   His way of treating them reminds us rather of the old fable of Proteus, who would transform himself into a whirl-

que des siens propres ; et quant à lui, s'il a accepté, avec une sincère satisfaction, les honneurs scientifiques qui lui ont été prodigués à si juste titre, il a constamment refusé toutes les autres distinctions et les récompenses qu'on eût voulu lui décerner.   Il s'est contenté toute sa vie de la position relativement modeste qu'il occupait à l'Institution Royale de Londres ; avoir son laboratoire et strictement de quoi vivre, c'est tout ce qu'il lui fallait.—Presinge, le 29 août, 1867.—A. DE LA RIVE."

wind or a dragon, a flame of fire, or a rushing stream, in order to elude his pursuer ; but if the wary inquirer could catch him asleep in his cave, he might be constrained to utter all his secret knowledge : for the favourite thought of Faraday seems to have been that these various forces were the changing forms of a Proteus, and his great desire seems to have been to learn the secret of their origin and their transformations.   Thus he loved to break down the walls of separation between different classes of phenomena, and his eye doubtless sparkled with delight when he saw what had always been looked upon as permanent gases liquefy like common vapours under the constraint of pressure and cold —when the wires that coiled round his magnets gave signs of an electric wave, or coruscated with sparks—when the electricities derived from the friction machine and from the voltaic pile yielded him the same series of phenomena— when he recognized the cumulative proof that the quantity of electricity in a galvanic battery is exactly proportional to the chemical action—when his electro-static theory seemed to break down the barrier between conductors and insula- tors, and many other barriers beside—when he sent a ray of polarized light through a piece of heavy glass between the poles of an electro-magnet, and on making contact saw that the plane of polarization was rotated, or, as he said, the light was magnetized—and when he watched pieces of bismuth, or crystals of Iceland spar, or bubbles of oxygen, ranging themselves in a definite position in the magnetic field.

" I delight in hearing of exact numbers, and the determi- nations of the equivalents of force when different forms of force are compared one with another," he wrote to Joule in 1845 ; and no wonder, for these quantitative comparisons have proved many of his speculations to be true, and have made them the creed of the scientific world.   When he began to investigate the different sciences, they might be compared to so many separate countries with impassable frontiers, different languages and laws, and various weights and measures ; but when he ceased they resembled rather a brotherhood of states, linked together by a community of interests and of speech, and a federal code ; and in bringing about this unification no one had so great a share as himself.

He loved to speculate, too, on Matter and Force, on the nature of atoms and of imponderable agents.   " It is these things," says the great German physicist Professor Helmholz, " that Faraday in his mature works ever seeks to purify more and more from everything that is theoretical, and is not the direct and simple expression of the fact.   For instance, he contended against the action of forces at a distance, and the adoption of two electrical and two magnetic fluids, as well as all hypotheses contrary to the law of the conservation of force, which he early foresaw, though he misunderstood it in its scientific expression.   And it is just in this direction that he exercised the most unmistakeable influence first of all on the English physicists." [1]

While, however, Faraday was pre-eminently an experimental philosopher, he was far from being indifferent to the useful applications of science.   His own connection with the practical side of the question was threefold : he undertook some laborious investigations of this nature himself; he was

[1] Preface to " Faraday und seine Entdeckungen."

frequently called upon, especially by the Trinity House, to give
his opinions on the inventions of others ; and he was fond of
bringing useful inventions before the members of the Royal
Institution in his Friday evening discourses.   The first of
these, on February 3, 1826, was on India-rubber, and was
illustrated by an abundance of specimens both in the raw
and manufactured states.   In this way also he continued to
throw the magic of his genius around Morden's machinery
for manufacturing Bramah's locks, Ericsson's caloric engine,
Brunel's block machinery at Portsmouth, Petitjean's process
for silvering mirrors, the prevention of dry-rot in timber,
De la Rue's envelope machinery, artificial rubies, Bonelli's
electric silk loom, Barry's mode of ventilating the House of
Lords, and many kindred subjects.

It may not be amiss to describe the last of his Friday
evenings, in which he brought before the public Mr. C. W.
Siemens' Regenerative Gas Furnace.   The following letter
to the inventor will tell the first steps :—

"ROYAL INSTITUTION, *March* 22, 1862.
"MY DEAR SIR,

"I have just returned from Birmingham—and there
saw at Chance's works the application of your furnaces
to glass-making.   I was very much struck with the whole
matter.

"As our managers want me to end the F. evenings
here after Easter, I have looked about for a thought, for I
have none in myself.   I think I should like to speak of the
effects I saw at Chance's, if you do not object.   If you
assent, can you help me with any drawings or models, or

illustrations either in the way of thoughts or experiments?
Do not say much about it out of doors as yet, for my mind
is not settled in what way (if you assent) I shall present the
subject.

"Ever truly yours,

"C. W. SIEMENS, ESQ."   "M. FARADAY.

Of course the permission was gladly given, and Mr.
Siemens met him at Birmingham, and for two days con-
ducted him about works for flint and crown glass, or for
enamel, as well as about ironworks, in which his principle
was adopted, wondering at the Professor's simplicity of
character as well as at his ready power of grasping the
whole idea.   Then came the Friday evening, 20th June,
1862, in which he explained the great saving of heat
effected, and pictured the world of flame into which he had
gazed in some of those furnaces.   But his powers of lecturing were enfeebled, and during the course of the hour he
burnt his notes by accident, and at the conclusion he very
pathetically bade his audience farewell, telling them that he
felt he had been before them too long, and that the experience of that evening showed he was now useless as their
public servant, but he would still endeavour to do what he
could privately for the Institution.   The usual abstract of
the lecture appeared, but not from his unaided pen.

Inventors, and promoters of useful inventions, frequently
benefited by the advice of Faraday, or by his generous help.
A remarkable instance of this was told me by Cyrus Field.
Near the commencement of his great enterprise, when he
wished to unite the old and the new worlds by the

telegraphic cable, he sought the advice of the great electrician, and Faraday told him that he doubted the possibility of getting a message across the Atlantic.   Mr. Field saw that this fatal objection must be settled at once, and begged Faraday to make the necessary experiments, offering to pay him properly for his services.   The philosopher, however, declined all remuneration, but worked away at the question, and presently reported to Mr. Field :—" It can be done, but you will not get an instantaneous message."   " How long will it take? " was the next inquiry.   "Oh, perhaps a second."   " Well, that's quick enough for me," was the conclusion of the American ; and the enterprise was proceeded with.

As to the electric telegraph itself, Faraday does not appear among those who claim its parentage, but he was constantly associated with those who do ; his criticisms led Ritchie to develop more fully his early conception, and he was constantly engaged with batteries and wires and magnets, while the telegraph was being perfected by others, and especially by his friend Wheatstone, whose name will always be associated with what is perhaps the most wonderful invention of modern times.

As to Faraday's own work in applied science, his attempts to improve the manufacture of steel, and afterwards of glass for optical purposes, were among the least satisfactory of his researches.   He was more successful in the matter of ventilation of lamp-burners.   The windows of lighthouses were frequently found streaming with water that arose from the combustion of the oil, and in winter this was often converted into thick ice.   He devised a plan by which this water was effectually carried away, and the room was also made more healthy for the keepers. At the Athenæum Club serious complaints were made that the brilliantly lighted drawing-room became excessively hot, and that headaches were very common, while the bindings of the books were greatly injured by the sulphuric acid that arose from the burnt coal-gas. Faraday cured this by an arrangement of glass cylinders over the ordinary lamp chimneys, and descending tubes which carried off the whole products of combustion without their ever mixing with the air of the room. This principle could of course be applied to brackets or chandeliers elsewhere, but the Professor made over any pecuniary benefit that might accrue from it to his brother, who was a lamp manufacturer and had aided him in the invention.

The achievements of Faraday are certainly not to be tested by a money standard, nor by their immediate adaptation to the necessities or conveniences of life. " Practical men " might be disposed to think slightly of the grand discoveries of the philosopher. Their ideas of "utility" will probably be different. One man may take his wheat corn and convert it into loaves of bread, while his neighbour appears to lose his labour by throwing the precious grain into the earth : but which is after all most productive? The loaves will at once feed the hungry, but the sower's toil will be crowned in process of time by waving harvests.

Yet some of Faraday's most recondite inquiries did bear practical fruit even during his own lifetime. In proof of this I will take one of his chemical and two of his electrical discoveries.

Long ago there was a Portable Gas Company, which made oil-gas and condensed it into a liquid. This liquid Faraday examined in 1824, and he found the most important constituent of it to be a light volatile oil, which he called bicarburet of hydrogen. The gas company, I presume, came to an end; but what of the volatile liquid? Obtained from coal-tar, and renamed Benzine or Benzol, it is now prepared on a large scale, and used as a solvent in some of our industrial arts. But other chemists have worked upon it, and torturing it with nitric acid, they have produced nitrobenzol—a gift to the confectioner and the perfumer. And by attacking this with reducing agents there was called into existence the wondrous base aniline,—wondrous indeed when we consider the transformations it underwent in the hands of Hofmann, and the light it was made to throw on the internal structure of organic compounds. Faraday used sometimes to pay a visit to the Royal College of Chemistry, and revel in watching these marvellous reactions. But aniline was of use to others besides the theoretical chemist. Tortured by fresh appliances, this base gave highly-coloured bodies which it was found possible to fix on cotton as well as woollen and silken fabrics, and thence sprang up a large and novel branch of industry, while our eyes were delighted with the rich hues of mauve and magenta, the Bleu de Paris, and various other " aniline dyes."

Everyone who is at all acquainted with the habits of electricity knows that the most impassable of obstacles is the air, while iron bolts and bars only help it in its flight : yet, if an electrified body be brought near another body, with this invisible barrier between them, the electrical state of the second body is disturbed. Faraday thought much over this question of "induction," as it is called, and found himself greatly puzzled to comprehend how a body should act where it is not. At length he satisfied himself by experiment that the interposed obstacle is itself affected by the electricity, and acquires an electro-polar state by which it modifies electric action in its neighbourhood. The amount varies with the nature of the substance, and Faraday estimated it for such dielectrics as sulphur, shellac, or spermaceti, compared with air. He termed this new property of matter "specific inductive capacity," and figured in his own mind the play of the molecules as they propagated and for a while retained the force. Now, these very recondite observations were opposed to the philosophy of the day, and they were not received by some of the leading electricians, especially of the Continent, while those who first tried to extend his experiments blundered over the matter. However, the present Professor Sir William Thomson, then a student at Cambridge, showed that while Faraday's views were rigorously deducible from Coulomb's theory, this discovery was a great advance in the philosophy of the subject. When submarine telegraph wires had to be manufactured, Thomson took "specific inductive capacity" into account in determining the dimensions of the cable : for we have there all the necessary conditions—the copper wire is charged with electricity, the covering of gutta-percha is a "dielectric," and the water outside is ready to have an opposite electric condition induced in it. The result is that, as Faraday himself predicted, the message is somewhat retarded ; and of course

it becomes a thing of importance so to arrange matters that
this retardation may be as small as possible, and the signals
may follow one another speedily.   Now this must depend
not only on the thickness of the covering, but also on the
nature of the substance employed, and it was likely enough
that gutta-percha was not the best possible substance.   In
fact, when Professor Fleeming Jenkin came to try the
inductive capacity of gutta-percha by means of the Red Sea
cable, he found it to be almost double that of shellac, which
was the highest that Faraday had determined, and attempts
have been made since to obtain some substance which should
have less of this objectionable quality and be as well
adapted otherwise for coating a wire.   There is Hooper's
material, the great merit of which is its low specific inductive capacity, so that it permits of the sending of four
signals while gutta-percha will only allow three to pass
along; and Mr. Willoughby Smith has made an improved
kind of gutta-percha with reduced capacity.   Of course no
opinion is expressed here on the value of these inventions,
as many other circumstances must be taken into account,
such as their durability and their power of insulation,—that
is, preventing the leakage of the galvanic charge; but at
least they show that one of the most abstruse discoveries of
Faraday has penetrated already into our patent offices and
manufactories.   Two students in the Physical Laboratory
at Glasgow have lately determined with great care the
inductive capacity of paraffin, and there can be little doubt
that the speculations of the philosopher as to the condition
of a dielectric will result in rendering it still more easy than
at present to send words of information or of friendly

greeting to our cousins across the Atlantic or the Indian Ocean.

The history of the magneto-electric light affords another remarkable instance of the way in which one of Faraday's most recondite discoveries bore fruit in his own lifetime; and it is the more interesting as it fell to his own lot to assist in bringing the fruit to maturity.

"BRIGHTON, *November* 29, 1831.

"DEAR PHILLIPS,

"For once in my life I am able to sit down and write to you without feeling that my time is so little that my letter must of necessity be a short one; and accordingly I have taken an extra large sheet of paper, intending to fill it with news.

"But how are you getting on? Are you comfortable? And how does Mrs. Phillips do; and the girls? Bad correspondent as I am, I think you owe me a letter; and as in the course of half an hour you will be doubly in my debt, pray write us, and let us know all about you. Mrs. Faraday wishes me not to forget to put her kind remembrances to you and Mrs. Phillips in my letter. . . . .

"We are here to refresh. I have been working and writing a paper that always knocks me up in health; but now I feel well again, and able to pursue my subject; and now I will tell you what it is about. The title will be, I think, 'Experimental Researches in Electricity':—I. On the Induction of Electric Currents; II. On the Evolution of Electricity from Magnetism; III. On a new Electrical Condition of Matter; IV. On Arago's Magnetic Phenomena.

There is a bill of fare for you; and, what is more, I hope
it will not disappoint you. Now, the pith of all this I
must give you very briefly; the demonstrations you shall
have in the paper when printed. . . . ."

So wrote Faraday to his intimate friend Richard Phillips,
on November 29th, 1831, and the letter goes on to describe
the great harvest of results which he had gathered since
the 29th of August, when he first obtained evidence of an
electric current from a magnet. A few days afterwards he
was at work again on these curious relations of magnetism
and electricity in his laboratory, and at the Round Pond
in Kensington Gardens, and with Father Thames at
Waterloo Bridge. On the 8th of February he entered in
his note-book : " This evening, at Woolwich, experimented
with magnet, and for the first time got the magnetic spark
myself. Connected ends of a helix into two general ends,
and then crossed the wires in such a way that a blow at $a\,b$
would open them a little. Then bringing $a\,b$ against the
poles of a magnet, the ends were disjoined, and bright
sparks resulted."

Next day he repeated this experiment at home with Mr.
Daniell's magnet, and then invited some of his best friends
to come and see the tiny speck of light.[1]

[1] I am indebted to Sir Charles Wheatstone for the following im-
promptu by Herbert Mayo :—

> " Around the magnet Faraday
> Was sure that Volta's lightnings play :
>   But how to draw them from the wire ?
> He drew a lesson from the heart :
> 'Tis when we meet, 'tis when we part,
>   Breaks forth the electric fire."

But what was the use of this little spark between the shaken wires? "What is the use of an infant?" asked Franklin once, when some such question was proposed to him. Faraday said that the experimentalist's answer was, "Endeavour to make it useful." But he passed to other researches in the same field.

"I have rather been desirous," he says, "of discovering new facts and new relations dependent on magneto-electric induction, than of exalting the force of those already obtained; being assured that the latter would find their full development hereafter." And in this assurance he was not mistaken. Electro-magnetism has been taken advantage of on a large scale by the metallurgist and the telegrapher; and even the photographer and sugar-refiner have attempted to make it their servant; but it is its application as a source of light that is most interesting to us in connection with its discoverer.

Many "electric lights" were invented by "practical men," the power being generally derived from a galvanic battery; and it was discovered that by making the terminals of the wires of charcoal, the brilliancy of the spark could be enormously increased. Some of these inventions were proposed for lighthouses, and so came officially under the notice of Faraday as scientific adviser to the Trinity House. Thus he was engaged in 1853 and 1854 with the beautiful electric light of Dr. Watson, which he examined most carefully, evidently hoping it might be of service, and at length he wrote an elaborate report pointing out its advantages, but at the same time the difficulties in the way of its practical adoption. The Trinity Corporation passed a special vote of thanks for his report, and hesitated to proceed further in the matter.

But Faraday's own spark was destined to be more successful. In 1853 some large magneto-electric machines were set up in Paris for producing combustible gas by the decomposition of water. The scheme failed, but a Mr. F. H. Holmes suggested that these expensive toys might be turned to account for the production of light. " My propositions," he told the Royal Commissioners of Lighthouses, " were entirely ridiculed, and the consequence was, that instead of saying that I thought I could do it, I promised to do it by a certain day. On that day, with one of Duboscq's regulators or lamps, I produced the magneto-electric light for the first time ; but as the machines were ill-constructed for the purpose, and as I had considerable difficulty to make even a temporary adjustment to produce a fitting current, the light could only be exhibited for a few minutes at a time." He turned his attention to the reconstruction of the machines, and after carrying on his experiments in Belgium, he applied to the Trinity Board in February 1857. Here was the tiny spark, which Faraday had produced just twenty-five years before, exalted into a magnificent star, and for Faraday it was reserved to decide whether this star should shed its brilliance from the cliffs of Albion. A good piece of optical apparatus, intended for the Bishop Rock in the Scillies, happened to be at the experimental station at Blackwall, and with this comparative experiments were made. We can imagine something of the interest with which Faraday watched the light from Woolwich, and asked questions of the inventor about all the

details of its working and expense; and we can picture the
alternations of hope and caution as he wrote in his report,
"The light is so intense, so abundant, so concentrated and
focal, so free from under-shadows (caused in the common
lamp by the burner), so free from flickering, that one cannot
but desire it should succeed. But," he adds, "it would
require *very careful* and progressive introduction—men with
peculiar knowledge and skill to attend it; and the means of
instantly substituting one lamp for another in case of accident. The common lamp is so simple, both in principle
and practice, that its liability to failure is very small. There
is no doubt that the magneto-electric lamp involves a great
number of circumstances tending to make its application
more refined and delicate; but I would fain hope that none
of these will prove a barrier to its introduction. Nevertheless, it must pass into practice only through the ordeal of a
full, searching, and prolonged trial." This trial was made
in the upper of the two light towers at the South Foreland;
but it was not till the 8th December, 1858, that the experiment was commenced. Faraday made observations on it
for the first two days, but it did not act well, and was discontinued till March 28, 1859, when it again shot forth its
powerful rays across the Channel.

It was soon inspected by Faraday inside and outside, by
land and by sea. His notes terminate in this way:—"Went
to the hills round, about a mile off, or perhaps more, so as
to see both upper and lower light at once. The effect was
very fine. The lower light does not come near the upper
in its power, and, as to colour, looks red whilst the upper
is white. The visible rays proceed from both horizontally,

but those from the low light are not half so long as those from the electric light. The radiation from the upper light was beautifully horizontal, going out right and left with intenseness like a horizontal flood of light, with blackness above and blackness below, yet the sky was clear and the stars shining brightly. It seemed as if the lanthorn [1] only were above the earth, so dark was the part immediately below the lanthorn, yet the whole tower was visible from the place. As to the shadows of the uprights, one could walk into one and across, and see the diminution of the light, and could easily see when the edge of the shadow was passed. They varied in width according to the distance from the lanthorn. With upright bars their effect is considerable at a distance, as seen last night; but inclining these bars would help in the distance, though not so much as with a light having considerable upright dimension, as is the case with an oil-lamp.

"The shadows on a white card were very clear on the edge—a watch very distinct and legible. On lowering the head near certain valleys, the feeble shadow of the distant grass and leaves was evident. The light was beautifully steady and bright, with no signs of variation—the appearance was such as to give confidence to the mind—no doubt about its continuance.

"As a light it is unexceptionable—as a magneto-electric light wonderful—and seems to have all the adjustments of quality and more than can be applied to a voltaic electric light or a Ruhmkorff coil."

---

[1] The room with glass sides, from which the light is exhibited at the top of a lighthouse, is called by this name.

The Royal Commissioners and others saw with gratification this beautiful light, and arrangements were made for getting systematic observations of it by the keepers of all the lighthouses within view, the masters of the light-vessels that guard the Goodwin Sands, and the crews of pilot cutters; after which Faraday wrote a very favourable report, saying, among other things: " I beg to state that in my opinion Professor Holmes has practically established the fitness and sufficiency of the magneto-electric light for lighthouse purposes, so far as its nature and management are concerned. The light produced is powerful beyond any other that I have yet seen so applied, and in principle may be accumulated to any degree; its regularity in the lanthorn is great, its management easy, and its care there may be confided to attentive keepers of the ordinary degree of intellect and knowledge." [1]

The Elder Brethren then wished a further trial of six months, during which time the light was to be entirely under their own control. It was therefore again kindled on August 22, and the experiment happened soon to be exposed to a severe test, as one of the light-keepers, who had been accustomed to the arrangement of the lamps in the lantern, was suddenly removed, and another took his place without any previous instruction. This man thought the light sufficiently strong if he allowed the carbon points to touch, as the lamp then required no attendance whatever,

[1] One night there was a beautiful aurora. Mr. Holmes remarked that his poor electric light could not compare with that for beauty; but Faraday rejoined, "Don't abuse your light. The aurora is very beautiful, and so is a wild horse, but you have tamed it and made it valuable."

and he could leave it in that way for hours together. On
being remonstrated with, he said, " It is quite good enough."
Notwithstanding such difficulties as these, the experiment
was considered satisfactory, but it was discontinued at the
South Foreland, for the cliffs there are marked by a double
light, and the electric spark was so much brighter than the
oil flames in the other house, that there was no small danger
of its being seen alone in thick weather, and thus fatally
misleading some unfortunate vessel.

After this Faraday made further observations, estimates
of the expense, and experiments on the divergence of the
beam, while Mr. Holmes worked away at Northfleet per-
fecting his apparatus, and the authorities debated whether
it was to be exhibited again at the Start, which is a revolving
light, or at Dungeness, which is fixed. The scientific
adviser was in favour of the Start, but after an interview
with Mr. Milner Gibson, then President of the Board of
Trade, Dungeness was determined on ; a beautiful small
combination of lenses and prisms was made expressly for it
by Messrs. Chance, and at last, after two years' delay, the
light again shone on our southern coast.

It may be well to describe the apparatus. There are 120
permanent magnets, weighing about 50 lbs. each, ranged on
the periphery of two large wheels. A steam-engine of about
three-horse power causes a series of 180 soft iron cores,
surrounded by coils of wire, to rotate past the magnets.
This calls the power into action, and the small streams of
electricity are all collected together, and by what is called
a " commutator " the alternate positive and negative currents are brought into one direction. The whole power is

M

then conveyed by a thick wire from the engine-house to the
lighthouse tower, and up into the centre of the glass apparatus. There it passes between two charcoal points, and
produces an intensely brilliant continuous spark. At sunset
the machine is started, making about 100 revolutions per
minute; and the attendant has only to draw two bolts in
the lamp, when the power thus spun in the engine-room
bursts into light of full intensity. The "lamp" regulates
itself, so as to keep the points always at a proper distance
apart, and continues to burn, needing little or no attention
for three hours and a half, when, the charcoals being con-
sumed, the lamp must be changed, but this is done without
extinguishing the light.

Again there were inspections, and reports from pilots and
other observers, and Faraday propounded lists of questions
to the engineer about bolts and screws and donkey-engines,
while he estimated that at the Varne light-ship, about equi-
distant from Cape Grisnez and Dungeness, the maximum
effect of the revolving French light was equalled by the
constant gleam from the English tower. But delays again
ensued till intelligent keepers could be found and properly
instructed; but on the 6th June, 1862, Faraday's own light,
the baby grown into a giant, shone permanently on the coast
of Britain.

France, too, was alert. Berlioz's machine, which was
displayed at the International Exhibition in London, and
which was also examined by Faraday, was approved by the
French Government, and was soon illuminating the double
lighthouse near Havre. These magneto-electric lights on
either side of the Channel have stood the test of years; and

for the last twelvemonth there has shone another still more beautiful one at Souter Point, near Tynemouth ; while the narrow strait between England and France is now guarded by these " sentinels of peaceful progress," for the revolving light at Grisnez has been lately illuminated on this principle, and on the 1st of January of this year the two lights of the South Foreland flashed forth with the electric flame.[1]

In describing thus the valuable applications of Faraday's discoveries of benzol, of specific inductive capacity, and of magneto-electricity, it is not intended to exalt these above other discoveries which as yet have paid no tribute to the material wants of man. The good fruit borne by other researches may not be sufficiently mature, but it doubtless contains the seeds of many useful inventions. Yet, after all, we must not measure the worth of Faraday's discoveries by any standard of practical utility in the present or in the future. His chief merit is that he enlarged so much the boundaries of our knowledge of the physical forces, opened up so many new realms of thought, and won so many heights which have become the starting-points for other explorers.

[1] The illuminating apparatus at Dungeness is one of what is termed the sixth order, 300 millimetres (about 12 inches) in diameter. Mr. Chance constructed one for Souter Point of the third order, one metre (nearly 40 inches) in diameter, with special arrangements for giving artificial divergence to the beam in a vertical direction, in order to obviate the danger arising from the luminous point not being always precisely in the same spot. It has also additional contrivances for utilizing the back light. Similar arrangements have been made for the South Foreland lights, which are also of the third order ; and every portion of the machinery and apparatus is in duplicate in case of accident, and the double force can be employed in times of fog.

SUPPLEMENTARY PORTRAITS

Iт has been said that there is no photograph or painting of Faraday which is a satisfactory likeness; not because good portraits have never been published, but because they cannot give the varied and ever-shifting expression of his features. Similarly, I fear that the mental portraiture which I have attempted will fail to satisfy his intimate acquaintance. Yet, as one who never saw him in the flesh may gain a good idea of his personal appearance by comparing several pictures, so the reader may learn more of his intellectual and moral features by combining the several estimates which have been made by different minds. Earlier biographies have been already referred to, but my sketch may well be supplemented by an anonymous poem that appeared immediately after his death, and by the words of two of the most distinguished foreign philosophers—Messrs. De la Rive and Dumas.

"Statesmen and soldiers, authors, artists,—still
   The topmost leaves fall off our English oak :
Some in green summer's prime, some in the chill
   Of autumn-tide, some by late winter's stroke.

" Another leaf has dropped on that sere heap—
     One that hung highest ; earliest to invite
The golden kiss of morn, and last to keep
     The fire of eve—but still turned to the light.

" No soldier's, statesman's, poet's, painter's name
     Was this, thro' which is drawn Death's last black line ;
But one of rarer, if not loftier fame—
     A priest of Truth, who lived within her shrine.

" A priest of Truth : his office to expound
     Earth's mysteries to all who willed to hear—
Who in the book of science sought and found,
     With love, that knew all reverence, but no fear.

" A priest, who prayed as well as ministered :
     Who grasped the faith he preached, and held it fast :
Knowing the light he followed never stirred,
     Howe'er might drive the clouds thro' which it past.

" And if Truth's priest, servant of Science too,
     Whose work was wrought for love and not for gain :
Not one of those who serve but to ensue
     Their private profit : lordship to attain

" Over their lord, and bind him in green withes,
     For grinding at the mill 'neath rod and cord ;
Of the large grist that they may take their tithes—
     So some serve Science that call Science lord.

" One rule his life was fashioned to fulfil :
     That he who tends Truth's shrine, and does the hest
Of Science, with a humble, faithful will,
     The God of Truth and Knowledge serveth best.

" And from his humbleness what heights he won !
     By slow march of induction, pace on pace,
Scaling the peaks that seemed to strike the sun,
     Whence few can look, unblinded, in his face.

" Until he reached the stand which they that win
        A bird's-eye glance o'er Nature's realm may throw ;
    Whence the mind's ken by larger sweeps takes in
        What seems confusion, looked at from below.

" Till out of seeming chaos order grows,
        In ever-widening orbs of Law restrained,
    And the Creation's mighty music flows
        In perfect harmony, serene, sustained ;

" And from varieties of force and power,
        A larger unity, and larger still,
    Broadens to view, till in some breathless hour
        All force is known, grasped in a central Will,

" Thunder and light revealed as one same strength—
        Modes of the force that works at Nature's heart—
    And through the Universe's veinèd length
        Bids, wave on wave, mysterious pulses dart.

" That cosmic heart-beat it was his to list,
        To trace those pulses in their ebb and flow
    Towards the fountain-head, where they subsist
        In form as yet not given e'en *him* to know.

" Yet, living face to face with these great laws,
        Great truths, great myst'ries, all who saw him near
    Knew him for child-like, simple, free from flaws
        Of temper, full of love that casts out fear :

" Untired in charity, of cheer serene ;
        Not caring world's wealth or good word to earn ;
    Childhood's or manhood's ear content to win ;
        And still as glad to teach as meek to learn.

" Such lives are precious : not so much for all
        Of wider insight won where they have striven,
    As for the still small voice with which they call
        Along the beamy way from earth to heaven."

                            Punch, September 7, 1867.

The estimate of M. A. de la Rive is from a letter he addressed to Faraday himself :—

" I am grieved to hear that your brain is weary ; this has sometimes happened on former occasions, in consequence of your numerous and persevering labours, and you will bear in mind that a little rest is necessary to restore you. You possess that which best contributes to peace of mind and serenity of spirit—a full and perfect faith, a pure and tranquil conscience, filling your heart with the glorious hopes which the Gospel imparts. You have also the advantage of having always led a smooth and well-regulated life, free from ambition, and therefore exempt from all the anxieties and drawbacks which are inseparable from it. Honour has sought you in spite of yourself; you have known, without despising it, how to value it at its true worth. You have known how to gain the high esteem, and at the same time the affection, of all those acquainted with you.

" Moreover, thanks to the goodness of God, you have not suffered any of those family misfortunes which crush one's life. You should, therefore, watch the approach of old age without fear and without bitterness, having the comforting feeling that the wonders which you have been able to decipher in the book of nature must contribute to the greater reverence and adoration of their Supreme Author.

" Such, my dear friend, is the impression that your beautiful life always leaves upon me ; and when I compare it with our troubled and ill-fulfilled life-course, with all that accumulation of drawbacks and griefs by which mine in particular has been attended, I put you down as very

happy, especially as you are worthy of your good fortune.
This leads me to reflect on the miserable state of those who
are without that religious faith which you possess in so great
a degree."

In M. Dumas' Eloge at the Académie des Sciences, occur
the following sentences :—

"I do not know whether there is a *savant* who would
not feel happy in leaving behind him such works as those
with which Faraday has gladdened his contemporaries, and
which he has left as a legacy to posterity : but I am certain
that all those who have known him would wish to approach
that moral perfection which he attained to without effort.
In him it appeared to be a natural grace, which made him
a professor full of ardour for the diffusion of truth, an
indefatigable worker, full of enthusiasm and sprightliness
in his laboratory, the best and most amiable of men in the
bosom of his family, and the most enlightened preacher
amongst the humble flock whose faith he followed.

"The simplicity of his heart, his candour, his ardent
love of the truth, his fellow-interest in all the successes,
and ingenuous admiration of all the discoveries of others,
his natural modesty in regard to what he himself discovered,
his noble soul—independent and bold,—all these combined
gave an incomparable charm to the features of the illustrious physicist.

"I have never known a man more worthy of being
loved, of being admired, of being mourned.

"Fidelity to his religious faith, and the constant observance of the moral law, constitute the ruling characteristics
of his life.   Doubtless his firm belief in that justice on high

which weighs all our merits, in that sovereign goodness which weighs all our sufferings, did not inspire Faraday with his great discoveries, but it gave him the straight-forwardness, the self-respect, the self-control, and the spirit of justice, which enabled him to combat evil fortune with boldness, and to accept prosperity without being puffed up. . . .

"There was nothing dramatic in the life of Faraday. It should be presented under that simplicity of aspect which is the grandeur of it. There is, however, more than one useful lesson to be learnt from the proper study of this illustrious man, whose youth endured poverty with dignity, whose mature age bore honours with moderation, and whose last years have just passed gently away surrounded by marks of respect and tender affection."

APPENDIX.

1835. Corresponding member of the Royal Academy of Medicine, Paris.

Honorary member of the Royal Society, Edinburgh.

Honorary member of the Institution of British Architects.

Honorary member of the Physical Society, Frankfort.

Honorary Fellow of the Medico-Chirurgical Society, London.

1836. Senator of the University of London.

Honorary member of the Society of Pharmacy, Lisbon.

Honorary member of the Sussex Royal Institution.

Foreign member of the Society of Sciences, Modena.

Foreign member of the Natural History Society, Basle.

1837. Honorary member of the Literary and Scientific Institution, Liverpool.

1838. Honorary member of the Institution of Civil Engineers.

Foreign member of the Royal Academy of Sciences, Stockholm.

1840. Member of the American Philosophical Society, Philadelphia.

Honorary member of the Hunterian Medical Society, Edinburgh.

1842. Foreign Associate of the Royal Academy of Sciences, Berlin.

1843. Honorary member of the Literary and Philosophical Society, Manchester.

Honorary member of the Useful Knowledge Society, Aix-la-Chapelle.

1844. Foreign Associate of the Academy of Sciences, Paris.

Honorary member of the Sheffield Scientific Society.

1845. Corresponding member of the National Institute, Washington.

Corresponding member of the Société d'Encouragement, Paris.

1846. Honorary member of the Society of Sciences, Vaud.

1847. Member of the Academy of Sciences, Bologna.

Foreign Associate of the Royal Academy of Sciences of Belgium.

Fellow of the Royal Bavarian Academy of Sciences, Munich.

Correspondent of the Academy of Natural Sciences, Philadelphia.

1848. Foreign honorary member of the Imperial Academy of Sciences, Vienna.

1849. Honorary member, first class, of the Institut Royal des Pays-Bas.

Foreign correspondent of the Institute, Madrid.

1850. Corresponding Associate of the Accademia Pontificia, Rome.

Foreign Associate of the Academy of Sciences, Haarlem.

APPENDIX.

1851. Member of the Royal Academy of Sciences, The Hague.
      Corresponding member of the Batavian Society of Experimental
        Philosophy, Rotterdam.
      Fellow of the Royal Society of Sciences, Upsala.
1853. Foreign Associate of the Royal Academy of Sciences, Turin.
      Honorary member of the Royal Society of Arts and Sciences,
        Mauritius.
1854. Corresponding Associate of the Royal Academy of Sciences,
        Naples.
1855. Honorary member of the Imperial Society of Naturalists,
        Moscow.
      Corresponding Associate of the Imperial Institute of Sciences
        of Lombardy.
1856. Corresponding member of the Netherlands' Society of Sciences,
        Batavia.
      Member of the Imperial Royal Institute, Padua.
1857. Member of the Institute of Breslau.
      Corresponding Associate of the Institute of Sciences, Venice.
      Member of the Imperial Academy, Breslau.
1858. Corresponding member of the Hungarian Academy of Sciences,
        Pesth.
1860. Foreign Associate of the Academy of Sciences, Pesth.
      Honorary member of the Philosophical Society, Glasgow.
1861. Honorary member of the Medical Society, Edinburgh.
1863. Foreign Associate of the Imperial Academy of Medicine, Paris.
1864. Foreign Associate of the Royal Academy of Sciences, Naples.

INDEX.

THE END.

W. CLAY, SONS, AND TAYLOR, PRINTERS, BREAD STREET HILL.

Printed in the United States
By Bookmasters